Science and the Future of Man

Science and the Future of Man is based on the proceedings of the conference sponsored by Boston College and the American Association for the Advancement of Science, December 28-29, 1969.

Science and the Future of Man

Editors
Robert L. Carovillano
James W. Skehan, S.J.

The MIT Press
Cambridge, Massachusetts, and London, England

501
C293s

To Our Concerned Colleagues

Contents

Contributors
ix

Preface
xi

Part I.
Science and the Problems of Society

The University Scientist
and Society   J. Tuzo Wilson
3

Scientists in the Courtroom   Victor J. Yannacone, jr.
20

Urban Problems and Technology Paul Parks
36

The Government and Science—
Troubled Waters   Donald F. Hornig
50

Discussion Session
57

Part II.
The Scientist and Society

Rethinking Scientific Objectives   Franklin A. Long
72

Science and the University Crisis John Platt
79

Problems and Responsibilities   George Wald
91

Why Basic Research?   Victor F. Weisskopf
96

The Integration of Science
and Life   Lewis Mumford
103

Discussion Session
113

**Part III.
Confrontation**

Science and the Revolution in the
Third World   Robert F. Drinan, S.J.
134

Crisis of Man and his
Environment   Edmund S. Muskie
147

Remarks   Philip H. Abelson
159

Remarks   Erwin D. Canham
163

Discussion Session
171

Name Index
189

Subject Index
191

Contributors

Philip H. Abelson
Director, Carnegie Geophysical Institute, and Editor of *Science*

Erwin D. Canham
Editor, *Christian Science Monitor*

Robert L. Carovillano
Chairman, Department of Physics,
Boston College

Robert F. Drinan, S.J.
Vice President and Provost,
Boston College

Donald F. Hornig
President, Brown University

W. Seavey Joyce, S.J.
President, Boston College

Franklin A. Long
Henry Luce Professor of Science and Society,
Cornell University

Lewis Mumford
Honorary Fellow:
Stanford University

Edmund S. Muskie
United States Senator, Maine

Paul Parks
Director, Model Cities Program, Boston

John Platt
Associate Director, Mental Health Research Institute,
The University of Michigan

James W. Skehan, S.J.
Director, Boston College Environmental Center
Boston College

George Wald
Professor of Biology,
Harvard University

Victor F. Weisskopf
Chairman of the Physics Department and Institute Professor, Massachusetts Institute of Technology

J. Tuzo Wilson
Principal, Erindale College,
University of Toronto

Victor J. Yannacone, jr.
Environmental Law Committee,
American Trial Lawyers Association

Preface

About one year ago a small group of senior scientists met with the graduate school dean, Samuel Aronoff, at Boston College and expressed their concern with the trends and state of science. Before long it was decided that a symposium should be planned and presented to an appropriate audience. The American Association for the Advancement of Science was holding its annual meeting in our fair city, Boston, on December 26-31, 1969, and the jointly sponsored program "Science and the Future of Man" emerged. The following synopsis was used in our program announcements:

The war in Viet Nam and changes in budgetary priority have acted in concert to cause the scientist to reappraise his career and discipline objectives. The dilemma of the socially-concerned and scientifically-oriented student introduces an important ramification of this consideration. The symposium "Science and the Future of Man" will analyze the role of the scientist and science in society. The contributions of science—both positive and negative—to our technological society and its most urgent problems will be studied, and ways in which science and scientists can contribute to a better society will be sought. Perspectives will be historical, immediate, and long-range. *Historical:* Issues and events relative to the role scientists have played in society in the past will be reviewed. Topics will include the development and effects of modern medicines, communications, transportation, weapon systems, nuclear energy, government macrofunding of science, and the scientist's past role in public policy. *Immediate:* Discussion will center on the vital present-day needs of society, both national and international, both urban and rural; the influence of science on national and international politics; the activities of the scientist in the university, in government, in industry. *Long Range:* Attention will focus on questions for the future: What are the best scientific objectives for society now? Where is society going? How can the scientist as an individual and as a member of a scientific community become a more effective and constructive force in society? How can environmental research be applied to urban problems, overpopulation, genetics, hunger, poverty? How can society be made aware of scientific developments directly affecting its future?

As the symposium actually developed, the approach to these problems became much more concrete, much more vital than we could have anticipated in our original projection. In fact, the excellent and profound treatment of the wide gamut of important and relevant topics concretized the foregoing future-oriented description of the program announcement. Moreover, important insights into other aspects not elaborated in the speeches emerged in the discussion at the conclusion of each panel.

This volume is based upon the actual proceedings of our symposium and the format of the book follows that of the actual program. The three parts correspond to our three half-day sessions, which each had a panel of principal speakers and a discussion. The text was carefully prepared from tape recordings of the program. Each speaker that presented prepared remarks was given the opportunity to revise or expand the text of his talk for this publication. Few major changes were made in this process, but some deletions did remove interesting features that the live audience experienced. Thus, some of the levity of Robert Drinan is not recorded as this distinguished law scholar braced himself to "give hell" to a seemingly hostile scientific audience. And the indignation of Lewis Mumford at the president's science advisor who suggested that industry and business should not be overly inconvenienced for the sake of possible environmental security is not expressed herein. The discussion sessions are complete and accurately follow the live program. Unfortunately, despite all best efforts, many persons who asked questions or offered comments did not introduce themselves and are designated as unidentified in the text.

Very many credits must be given in completing such an effort as this one. My colleague, James W. Skehan, S.J., functioned ably as co-arranger for the symposium and, of course, as co-editor of this book. Dr. Walter Berl, Meeting Editor for the American Association for the Advancement of Science, graciously agreed to the worthiness of our planned symposium at the outset and gave much valuable advice and support. At Boston College, Reverend Francis X. Shea, S.J., Executive Vice President, and Reverend Charles F. Donovan, S.J., Dean of Faculties, similarly endorsed our efforts and made necessary resources available. We drew upon the talents of a knowledgeable Planning Committee consisting of professors (E. Brooks, J. Gilroy, D. MacLean, S.J., J. Maguire, D. Plocke, S.J.) and students (P. Byrne, J. Brabson, S. Helwick). Drs. Maguire and MacLean, S.J., made particularly notable contributions. The ladies on our staffs that capably performed so many necessary tasks include Mrs. Grace McCarthy, Mrs. Phyllis Mooers, Miss Lydia McCarthy and Mrs. Cynthia Cudworth, and we are grateful to them. The completeness and reliability of this book was greatly aided in having available from Mr. Zugby of AAAS an independent tape recording or our symposium. My administrative assistant, Mr. John Burke, deserves special plaudits for his diligence and initiative in assisting in many ways, unobtrusively, throughout this undertaking.

R. L. Carovillano
Boston College

Overleaf
left: Session III, Robert F. Drinan, S.J. right: Session II, George Wald
Session III. Edmund S. Muskie, W. Seavey Joyce, S.J., Robert F. Drinan, S.J.
Session I. James W. Skehan, S.J., Paul Parks, J. Tuzo Wilson
Session II. Lewis Mumford, Robert L. Carovillano, George Wald, Victor F. Weisskopf, Franklin A. Long

Part I
Science and the Problems of Society
Chairman: James W. Skehan, S.J.

Chairman:

If one were to ask any professional geologist anywhere in the world to name five of the most stimulating scientists of modern time, this list is virtually certain to include the name of Tuzo Wilson. Having received his A.M. degree at the University of Cambridge, 1932, and Ph.D. from Princeton, 1936, he has had a most productive research and teaching career as a professor of geophysics, University of Toronto. He is author of numerous research papers and books on structure and growth of continents, the theory of continental drift, and the geology of ocean basins. His provocative significant discoveries and theories have earned for him numerous medals and awards. Moreover, his ideas have stimulated a wealth of research worldwide to test the validity of his conclusions.

Having taken on new responsibilities as Principal of Erindale College of the University of Toronto in 1967, Professor Wilson applies his uncanny intuition and talents to develop relevance in the wider sphere within the University, as well as in the relationship of the University to the rest of society.

The University Scientist and Society
J. Tuzo Wilson

Five comparisons between scientists and others
For several decades science has been growing more quickly than the economy, and scientists have been multiplying faster than the population. We have become accustomed to the joys of untrammeled, exponential growth. Such periods of free expansion are so exhilarating that it is depressing to realize they cannot long continue.

This limitation is one which many scientists have yet to recognize and accept. Such obtuseness is unscientific and shows that training in science does not teach all scientists to be unbiased. They plead a special case for science, and this special pleading is so common a human frailty that it suggests comparing some characteristics of men engaged in science with those in other endeavors. I have chosen five human characteristics which seem to me to illustrate resemblances between scientists, businessmen, revolutionaries, the military, and the religious.

The trait of faith
The first is a confident assurance in the truth and sanctity of an all-embracing credo. All great faiths stem from the recognition by humans of some greater power than themselves. They want to find it and with it discover a mode of life on which they can pattern their behavior without question. A great faith is a reassuring thing.

It is obvious that such beliefs motivate revolutionaries and the religious. It is not popular to do so but I suggest that faith in Marxism may be as genuine and may have the same effect on a communist that belief in Christianity has on a Christian. In similar ways military men obtain comfort from patriotism and the businessmen reassurance from their confidence in free enterprise. Men usually learn their creeds in childhood and it is characteristic to hold them fervently. Nevertheless many faiths are contradictory, so they cannot possibly all be true. This combination of antithetical opinions deeply cherished leads to violent conflicts, and those between different religious sects, between separate nations, and between upstarts and the established have been the chief causes of war.

The fervor with which a belief is held is not a guide to whether it is good or bad. The fact that millions believe in particular religions, patriotisms, enterprises, or revolutionary causes does not confer morality on them all. Although great religions have produced splendid human ideals and inspired dedicated lives, other religions no less strongly held have led to passionate involvement with human sacrifice, slavery, and other evils. The very real benefits of free enterprise have not prevented that system from producing such excesses as child labor, sweatshops, cartels, and swindles of many kinds. Patriotism has often been an excuse for jingoism and oppression. It is the nobility of the creed not the enthusiasm it generates that determines morality.

I suggest that the same rules apply to science. The faith of a scientist lies in the search for truth, but the fact that scientists believe strongly that they should be free to pursue truth as they see it is not enough to establish that all science is moral. Although pure science, if it is the unadulterated search for truth, may have intrinsic merit, how many scientists can claim that their search is unselfish and sublime? Only the inherent nature of a research project, not just a passion to undertake it, can establish its value. From the point of view of morality I suggest that one who does research to develop a nerve gas must share the guilt of him who unleashes it. The fact that research is involved does nothing to raise its morality. As a retired soldier and a recipient of defense grants I am prepared to listen to arguments that circumstances may make both research and shooting necessary, but shooting (and getting shot at) demands more courage and if anything is more a moral issue than weapons research. It is recognition of this truth that has grated on the conscience of the atomic scientists.

Contrary to what a great many scientists would like to believe, I maintain that it is just as wrong to claim that the nature of science in some way makes it holy and that any pursuit of it deserves support as it would be to claim that all religions, all revolutions, all free enterprise, and all nationalisms are sacred.

The preference for simplicity
The second human trait which I think applies to all our several fields is the preference for making a short, sharp attack on a small but frequently inconsequential problem which is easily defined in preference to tackling broad but vaguer problems of greater significance.

The fondness of the military for believing that they can solve problems by a sudden, brief assault without thinking where this will lead or about the causes of wars is notorious. The village priest is more inclined to insist that his parishioners adhere to the outward and visible signs of their religion than to worry about social problems. Businessmen think that what is good for them or for General Motors must be good for the whole country. Revolutionaries feel that by the mere overthrow of one system and its replacement by another they will produce paradise. So it is with science. Scientists have a very strong tendency to prefer limited problems of a rigorous nature which they can solve relatively completely and precisely to those larger and diffuse ones which are all-important for human survival.

The love of tools
The third characteristic I would choose is man's well-known love of tools. The word tools is often a euphemism for weapons upon which the military and the revolutionary would spend any available sums of money, however vast. Religions have encrusted spiritual intangibles with elaborate overtones of idolatry, ritual, and sacred relics. Businessmen bolster their egos with material signs of wealth. It often seems strange to a scientist to observe the store a businessman may set on the quality of his car or his golf clubs, but scientists have tools of their own to love. Instruments are their passion. In science one does not rely entirely on other people's results. More scientists find it important to build some equipment and discover a little thing of one's own. To relate broad results together is to gamble with one's career and to be a mere teacher is considered of little consequence.

The love of the hierarchy for professional secrecy
The fourth similarity among members of these institutions is the tendency of the hierarchies to wish to wrap their discoveries in secrecy. In all professions it is popular to consider that only professionals can understand and judge the inner mysteries.

Since the lay public does not know the secrets they should be excluded from the process of making decisions.

This is dangerous. It is when priesthoods become too strong that religions go astray; it is when revolutionaries become estranged from their followers that they turn into dictators; it is when executives hide matters from shareholders that delinquency most frequently arises; war is proverbially too important a matter to be left to the generals; I think the same danger faces science.

Many scientists feel that science is so difficult a matter that the decisions as to what is good science and what science should be supported can only be made by scientists. So long as there was lots of money and these decisions involved only questions of relative technical excellence, this view was tenable, but today when the choice is whether some whole fields should be strongly supported and others not, and when the question of human survival is at stake, it is certainly important to ask the lay public to see that scientific hierarchy is not just pursuing fun for fun's sake.

That accusation has often been made by those outside of science and because I entered a new and interdisciplinary field I have been in a good position to observe this from within. Forty years ago my professor had a hard battle to introduce geophysics into a department of physics whose members considered that only classical physics and the new atomic investigations were proper and orthodox. Subsequently I have observed that not all departments of geology have welcomed geophysics, but some have regarded it as an upstart disturbing the more petrified of their concepts.

The failure of the establishment to train great innovators
The fifth common feature is in the nature of a saving grace. It is that in all human institutions rarely does the establishment train

the truly great leaders. Even most of the minor innovators are self-generating mutants arising to refresh and strengthen the breed. This failure by the establishment arises because it seeks to train people to operate existing systems and never seeks to destroy them by building new rivals. It is clear enough that great religious leaders and revolutionaries have generally been regarded as an anathema by those who preceded them. Neither Henry Ford nor Walter Chrysler was the typical organization man, and they certainly upset the carriage manufacturers and saddleries. The military are notorious for their lethargy in adopting new weapons and tactics, and many leaders of revolutions have demonstrated that untrained civilians may be better generals than opponents who spent years in staff college.

It is sad to reflect that the same is true in science. Many of the greatest discoveries have been made by men like Einstein who were not products of the establishment. Others, like Newton and Darwin, admittedly received sound educations but they certainly ignored prevailing opinions when they made their great contributions. Koestler claims that the reason Copernicus did not publish his ideas until his deathbed was for fear that other scholars would laugh at him, while Galileo suffered persecution in establishing them. It is sad to reflect that the safe road in science and the one which the vast majority of scientists pursue is to discover more details in the kind of science in which they were trained. It is far harder to explore the truly new.

So far I have labored the thesis that scientists are but human. The most important conclusion I can draw is the desirability of having lay opinion to help them make central decisions about the support of science. Let us hope that it will be more efficacious than the control which politicians supposedly exert over the military.

Some characteristics of science
Now I wish to turn from scientists to the nature of science itself.
If scientists are unexceptional, science is certainly not. In this it
differs from religions, wars, revolutions, and business, all of which
have existed for millennia without effecting any change comparable to that wrought in the past few decades by the impact
of science and advanced technology. Religion, historically the
guardian and interpreter of the mysteries and the miracles of the
universe, now sees its sovereignty challenged by science.

What else is it but a miracle to learn that the solid earth is not the
fixed center of the universe but only a tiny wandering planet. Is it
not a miracle, and extraordinary beyond belief, that a yellow gas
could combine with an inflammable metal to make common salt?
How unimaginable and how unexpected was the discovery that
some ordinary matter can decay with the generation of enormous
nuclear power and that matter is made of particles which can be
guided to produce the wonders of electronics, television and computers and thus make space travel possible. Science has made
the impossible real, the unbelievable true, and a belief in miracles
commonplace.

This exceptional power of science, combined with the all-too-
human and commonplace character of scientists makes it doubly
important that the control of science should not rest solely in
the hands of a scientific priesthood but that its operations should
be exposed to the widest possible scrutiny from many points of
view and be subject to open criticism by many able minds.

Scrutiny of science
If science is to be subject to open scrutiny what are the questions which the public should ask about it? Some which I believe pertinent, if controversial and difficult, include the following:

We have become accustomed to the rapid unfolding of a succession of scientific miracles. Is this rate of discovery likely to continue indefinitely?

Will future discoveries, even if great and numerous, continue to have the same impact and be as valuable as have been those of the past?

Is there a proper balance between the research efforts in various fields, and is the direction in which recent technology is taking us healthy or one which mankind can long pursue?

If some of the results of technology are unhealthy, can this trend be corrected by more scientific research? (If not, what else is needed?)

Is enough consideration being given to answering these questions?

Finally, I come to the subject with which I was particularly asked to deal. Are the universities keeping pace with these changes, giving them adequate consideration and changing their habits of teaching and subjects of research to suit the changing world?

I regret that the only honest answer to all these questions is no for the reasons, and with the reservations, which follow.

Will great scientific discoveries long continue?
My reason for believing that the number of truly great discoveries to be made may be limited is that either the universe has a finite and definite structure or that it is infinitely complex. In the first case one sees signs that we may be approaching, albeit very imperfectly, the limits of the general structure of what there is to know. Perhaps this feeling is merely a reflection of our own inadequacy, but if the universe is indeed infinitely complex how much of it can the human mind ever understand? There is cer-

tainly a limit to human capacity and thus in either case the rate of great discoveries will slow, although the details will multiply.

It also appears that in every discipline the one greatest discovery is the realization that that subject is governed by extraordinary behavior. That is the time of scientific revolution, and once recognized in any discipline, nothing as spectacular seems to occur again, however vast the wealth of detail remaining to be elucidated.

Professor R.P. Feynman expressed this possibility very well at the M.I.T. Centennial Symposium on the Future of Physics. The last sixty years, he pointed out, had seen a revolution in physics arising from three tremendous discoveries. They were special and general relativity and quantum mechanics. He could not imagine this rate continuing, as it would require fifty similar discoveries during the next thousand years. He concluded by recommending that ambitious students try some other field less deeply mined than nuclear physics.

I am prepared to admit that, particularly in the behavioral and social sciences, there are vital discoveries yet to be made, but I think it is not axiomatic that great discoveries will continue to be made in the conventional fields of the physical sciences at the same rate that they have been in the recent past.This does not limit the amount of scientific work to be done, for even if the universe is finite, we are far from understanding it, but great discoveries may become less frequent and most progress may become duller and more detailed in nature.

Is the value of scientific discovery diminishing?
The second question is more important. It is whether the results of scientific research may not be of diminishing value to mankind. For example, if our grandfathers took several weeks to sail across the Atlantic, our fathers several days to steam across, and if we fly across in a few hours, will it necessarily be an advantage for

our children to suppose that they could cross in minutes or that
our grandchildren could cross in seconds? Everybody can see
that this is ridiculous. Also everybody knows that it would be
more useful if we could commute to the office on time.

Some think and many would like to think that this pessimism
about the future of science is a delusion. If they believe me to be
unnecessarily glum, I would urge them to read several articles in
the issue of the journal of this society, Science, for November 28
and the remarks of the president of this association, Dr. Bentley
Glass, in the AAAS Bulletin for September last. Dr. Glass named
five other specific factors that he thought would limit the growth
of science. I will only touch on one, the evil side effects of technology. In this he included pollution and the exhaustion of natural resources.

I can realize the effects of this latter rather clearly because my
life already spans the period of greatest change and most rapid
consumption of resources. In 1913 I made a trip to Europe which
deeply impressed my juvenile mind. A carriage with two horses
took us to the station on unpaved roads. The train had gas lamps.
Servants were abundant. Men jingled gold sovereigns in their
pockets. I doubt if my mother needed a passport. Blériot had recently made the first flight across the English Channel. If the
memories of the older of us can encompass the changes which
embrace the growth of the automobile, airplane, nuclear and
electronics industries, will not the youngest here see as much? I
should like to think it will be as good, but it is obvious that the
breathtaking changes of the past fifty years must diminish. The
new generation faces the totally different problem of slowing
down the rate of change and of adjusting life to a steady plateau
with little growth.

Take the question of resources. It astonishes me that hardheaded
businessmen with memories as long as mine do not seem to be
able to realize the imminence of short supply. Surely the excite-

ment about discovering oil in the Arctic or under the stormy ocean off Newfoundland arises only because oil can no longer be found in Texas!

If the ever-accelerating demands for petroleum which have been so voracious for the past fifty years continue, all discoverable oil will be used up in the next fifty. Realistically, one can only say what an excellent prospect, otherwise the fumes would choke us. Only a few natural minerals, including iron, aluminum, coal, and building materials are in abundance. Most others are in short supply, and already many of them are being imported into the United States.

The vigorous growth of Japan along American lines is almost wholly dependent upon imports of fuel and metals. This is producing temporary wealth for countries like my own by exhausting our supplies, but the fact that Japan has to import the materials that form the basis of its economy demonstrates that few if any other of the less-developed countries can repeat the Japanese experiment. There are no adequate sources of the raw materials.

The effect of scientific research and technological development has created rapid expansion. It is clearly a delusion to believe that this can long continue. We face a forced deceleration in rate of growth and progress. The question of whether future research will help us adjust to that deserves consideration.

Is support of science properly balanced?
Enough thought has not been given to the question of whether there is an appropriate balance in the distribution of scientific effort.

Because the results of science are so formidable, most people respect and fear it and wish to leave it alone. They make the mistake of taking the same attitude toward scientists. This is wrong because it is science, not scientists, which is formidable. Scientists are just ordinary. As a result scientists have been allowed their own way with the guidance, or nonguidance, of science. To leave any subject to the control of its own hierarchy is bound to produce distortion, and science has become distorted.

To say this is not to impute any dishonesty, laziness, or lack of well-intentioned motives to scientists. On the contrary they are mostly hard working and dedicated. The distortion arises from two passive causes which they have scarcely noticed. One follows the parable of the common lands, that any well-organized group will always tend to get more than its share of the available benefits than others which are less organized, and the other is that scientists tend to choose some subjects for research over others for the reasons given early in this paper.

Thus well-organized groups such as departments of defense or the petroleum industry tend to get a lot of research done which interests them, while less strongly organized groups like disarmament or antipollution groups or those who look for underground water tend to get little help with research. Most Western countries spend a vast amount on armaments and a high percentage of that on research and development. They spend smaller but still significant amounts on foreign aid, but practically nothing on research into how best to make aid effective. As a result much of the money spent on aid is wasted. In spite of the fact that defense research has made nuclear annihilation probable in the near future, virtually nothing is spent on research which might make disarmament easier. The vested interests are all on one side.

The second tendency of selectivity in fields of research is well illustrated by the lopsided balance of basic research in universities. Most universities are divided into traditional departments, and each department corresponds to a professional group. These have convinced themselves that some teaching and research is orthodox and proper and they pursue it. Other interdisciplinary subjects are neglected.

Those graduates who prefer careers other than academic leave universities; the less practical tend to stay. These naturally propagate their own kind of impracticability and the whole trend is toward narrower specialization.

We have thus arrived at the ridiculous situation in Canada where most new doctors of philosophy have been trained in such esoteric aspects of the physical sciences that they cannot get jobs in the subjects in which they have been trained, while there is the greatest need for men trained in neglected subjects. Thus too many earth scientists are being trained in mineralogy, high-pressure silicate geochemistry, rock mechanics, and other specialties at the same time that the mining and petroleum industries rely for staff on immigrants trained along more practical lines. There is a great shortage of water geologists, and there is no one at all anxious to teach earth sciences in the schools.

These are all perfectly natural consequences that have distorted science and will continue to do so unless corrected.

Is enough thought being given to the direction of research?
Not only does this natural distortion of emphasis in research need to be corrected, but I believe that a much greater and more deliberate effort needs to be made to direct science. If not, mankind faces disaster either from nuclear war, poisoning by pollution, exhaustion of resources, or all three.

The arguments that one cannot decide which research is likely to be fruitful or that it would ruin basic research to interfere with it seem to me to be nonsense, because a great process of selection is already in action and has not been noticed.

The suggestion that spin-off can have benefits in other subjects which justify defense or lunar research is specious for it can as well be reversed. Could there not be benefits in the spin-off from extensive research into disarmament or into preservation of the environment? They might even produce a more efficacious defense than the present precarious balance of terror.

Most of the arguments we hear in defense of leaving science alone is special pleading by a hierarchy that has been left alone in secrecy too long. It is only natural that scientists should not wish to give up this power. That some scientists claim that to direct science would ruin it merely illustrates their blindness to the subtle forms of direction already operating. I suggest that the processes of decision in science should be made open.

This is not to attack science or scientists. On the contrary, scientists have been both diligent and honest. However, science is so powerful that it is only with its help that mankind can save themselves from disaster, and it is therefore axiomatic that it must be wisely directed.

The role of universities
Finally—the role of universities. As a university professor of twenty-five years standing and as one now concerned with administering a campus, I can say that I think that the universities have become out of date and out of touch with the times and I plead guilty to complicity in allowing and even at times encouraging this to happen.

With the best of intentions professors have made a mockery of what they most claim to defend—academic freedom. This freedom is quite undemocratic because it has excluded the majority of academics, the students, and has given license to the faculty. This has not taken the form of malfeasance, but of blindness and inertia. No one minds their pursuing some things which are out of date and some which are frivolities. The objection is that they have ignored things vital to the survival of mankind.

The reasons are precisely those which led to the distortion in scientific research and they have been as little noticed: the control by a closed hierarchy, the powerful influence of well-meaning but selfish interests, and the human tendencies mentioned early in this paper.

I don't blame the students whose attitude ranges from restlessness to uproar. The universities have not changed their habits enough either to keep pace with the change in their own role or to keep pace with the world outside.

The change in their own role is that whereas they were founded to train men for a few learned professions—the church, the law, medicine, and teaching—and they formerly taught only the academically inclined, they now teach everyone.

On the other hand, although the world has changed, universities in general still have the same departmental structure which they had at the beginning of the century. As a result they neglect whole interdisciplinary fields of the utmost importance. They have too many faculty teaching obsolete subjects and courses and too few considering the new. They try far too hard to train graduate students in their own narrow specialty, not recognizing that the ablest scientists will be the first to train themselves, and that it is a mistake to lead mediocre students, incapable of branching off on their own, into highly specialized culs-de-sac. As late as a generation ago few young people had seen much, and teaching was based upon forcing them to learn. Today, through television and travel, many students have seen more than their professors, and the problems they face are greater than were ours and they seem to know it. I believe they are more anxious to learn and are more aware of what is important to learn.

Fortunately the universities, if slow, are wise and are changing. Their present troubles are leading to corrections. I consider that the two most important are to open all the councils of university to students and lay members. Contrary to what many academics fear, this does not lead to student take-over. Offered responsibility, responsible students appear and are voted into power. Most students find committee meetings boring, and once they can see that the attempt is being made to operate a university on sensible lines they cease to attend and reduce their complaints.

The other correction, at least in new campuses where it is possible, is not to establish the standard departments but to obtain an infusion of fresh ideas by making different divisions. Thus, in languages there would be not French, German, Classics, and Spanish departments but perhaps departments of written languages, of spoken languages, and of literary criticism. In science I like microlife sciences, macrolife sciences, and the same divisions for physical sciences. The mere disturbance of the old structure leads to new contacts, infuses new ideas, and promotes interdisciplinary studies.

I feel that the knowledge of the world which television and easy travel have provided coupled with the peril of the times has given more young people a different and more mature attitude than that which we had. Only a noisy minority have turned to anarchy and escapism. The older generation or at least the more conservative of them are so much victims of the excesses of their own faiths, the narrowness of their specialities, their dislike of strangers, their confidence in a glowing future that they are not to be trusted to form university policy alone. Perhaps students will be no better, but the future is for them and not for us, and they deserve a say in how to cope with it.

While I deplore its excesses, I consider that student unrest has been justified. I consider it nonsense to claim that science should not be directed, because I believe that it suffers from having been too strongly directed, although most scientists were too blind to notice it.

Surely both universities and science are so valuable, and the direction which they pursue is so important, that it is unwise to leave their direction in the hands of a cloistered internal hierarchy. Open consideration of policy for both is a necessity.

Chairman:
Whatever is to be accomplished in a major way to improve the quality of the human environment must almost certainly be as a result of litigation or legislation. The legal career of the dynamic and eloquent Victor Yannacone from 1960 to the present is most aptly described as that of advocate—an advocate of people and their environment. His strong background in the sciences and a unique sense of timing and of the dramatic give him a unique capability for developing imaginative and pioneering approaches to litigation in the arena of pollution abatement and in saving for present and future generations the natural nonrenewable resources.

Scientists in the Courtroom
Victor J. Yannacone, jr.

Let me state that I believe that a lawyer is an advocate. Some lawyers advocate in the courtroom; some lawyers advocate in the classroom; some lawyers advocate in the halls of government; some lawyers advocate in the smoke-filled rooms behind the halls of government. But any lawyer who doesn't advocate is not really a lawyer, learned as he might be in the knowledge of what the law once was.

I believe that law is the framework of civilization, and litigation is the civilized answer to trial by combat. The courtroom is the arena, the lawyers are the champions and the rules of evidence are the articles of war. Litigation is no longer a game, and the courtroom is not a playground for dandy gentlemen. Litigation is a form of mortal combat, and the courtroom is the arena. Lawyers are no longer disinterested observers, exercising their wit and erudition before a disinterested judge. In every lawsuit someone must win and someone must lose, although on any given set of facts, the winner and loser might be different at different times in history or in the context of different civilizations, nevertheless, rest assured, a winner and a loser there will be.

Great industries will never lack for advocates. Government will never lack for advocates; political organizations will never lack for advocates; and the established institutions of the political, industrial, military power structure in their rape of our human and natural resources and their prostitution of the legal profession need no more advocates. People need advocates. People need champions, and our human and natural resources need protection.

During the spring of 1968, the alumni of Yale Law School, who claim among their numbers half the justices of the Supreme Court, 10 percent of the United States' law teachers and any number of distinguished attorneys, held a reunion. The intellectual theme of that reunion weekend was "Law and the Urban Crisis." Five prominent legal educators, deans of their respective law schools, and distinguished urban legal scholars in their own right, were invited to address the alumni on this urgent question. But, just as the proceedings were to begin, a group of black law students, together with members of New Haven's black coalition, entered the auditorium and began to address themselves to the all-white speakers' platform and the all-white alumni audience. "You don't understand the problem at all," they said. "The problem is not law and the urban crisis. Law is the urban crisis—whitey's law." If democracy is to survive, society's social problems are necessarily its legal problems. The time has come to treat society itself as if it were the lawyer's client. As Che Guevara so astutely observed, violent revolution is possible only when people have lost all hope in justice from the courts.

As long as people believe that procedural due process exists, violent revolution must fail. Unfortunately, it now appears that when we look to the law for answers to our social problems, we find that the law itself is the cause of many of the problems. It is *the law* which zones the housing patterns which lead to building too many highways for too many autos. It is *the law* that expropriates public property for private promoters. It is *the law* which permits environmental degradation. It is *the law* which asserts equal protection of *the law* for the corporate person, that fictional bastard child of *the law* cloaked by the court with all the God-given rights of a human being, but without a soul to save or tail to kick. It is *the law* which asserts that equal protection for the corporation and denies it to the poor, the black, the American Indian, the inarticulate, the politically weak or ineffective. It is *the law* which creates and sustains the tax system, that encourages overpopulation and penalizes those who would remain single or with few children. Always it is *the law*.

Many of our environmental problems stem from the misguided attempts of shortsighted ecological Neanderthals to control the uncontrollable. Pesticide abuse is a classic example. The indiscriminate use of broad spectrum, long-persistent pesticides such as DDT, dieldrin, endrin, aldrin, toxaphene, and heptachlor have so altered the ecology of agricultural ecological systems that new pest species have sprung up. Throughout the history of modern agrichemical methods, industry has ignored the value of integrated control techniques where specific chemical bullets are used to augment the armory of natural and biological controls.

Utilizing water resources for waste disposal is still another example of the utter futility of man's attempted environmental control. Oceans, rivers, lakes are just like any other sink—they have a finite capacity for waste, after which they back up. Moreover, they fight back—algae blooms quickly decay into sulfurous miasmas.

The attempt to control the air may have even more disastrous repercussions than the attempt to control the waters. The air is not a limitless sink in which we can pour countless tons of noxious gases and poisonous particulates. The atmosphere too

has a finite capacity, and not only are we reaching the limit today, but our high-speed air transportation system has begun to
alter our weather patterns and climatological cycles. Again, environmental control really appears to be a license from nature
with a fee yet to be paid.

If the law is to be of any help, pollution hazards must be evaluated in terms of equitable criteria: "Is the damage serious, permanent and irreparable?"

Perhaps the pollutants most dangerous to the community are
those insidious toxicants like radionuclides and certain chlori-
nated hydrocarbon pesticides constantly cycling throughout the
world's ecosystems in sublethal concentrations.

The latent effects of these toxicants are only beginning to be noticed, while the magnitude of the vested interests economically
committed to ignoring the hazards. makes them appear presently
insurmountable. However, the worst offenders in the process of
environmental degradation are not the ruthless entrepreneurs dedicated to wanton exploitation of our natural resources, the prof-
iteers and abusers of the public's air and water, but those short-
sighted, allegedly public-interest agencies such as the U.S. Department of Agriculture, the Army Corps of Engineers, the Atomic Energy Commission, and many state and regional development agencies. Their mission-oriented determinations preclude
any consideration of long-term ecological consequences.

The most we can hope to do is to manage the environment for
the highest and best use of all our natural resources for the
greatest good of the greatest number of people. But, the conservation movement has consistently failed to recognize the value
of our intangible, national, natural resource treasures when
faced with the so-called cost-benefit analyses of government
agencies. A Trout Unlimited chapter in Pennsylvania objected to
the potential destruction of a beautiful forested valley with a mag-

nificent trout stream at the bottom, by a highway that had a benefit value of maybe 12 or 13 million dollars. I pointed out that the conservationist should look at the valley and ask, what is its replacement cost?

But they kept telling me that you cannot value the replacement cost of a wooded slope, or stream, or a valley that took the glaciers eons to make. Nevertheless we finally forced the conservationists to look at the problem through the eyes of a poor country lawyer from New York and come up with dollars-and-cents answers. The answer is very simple. For years the conservation movement has been relying on the cheap, almost unvalued labor of the Mighty Hand of God. It may take a million years for glaciers to carve a valley, but we have unions; we have bulldozers; we have contractors. It is quite easy to determine the cost of digging any size valley by the cubic yard. So we figure how many billion cubic yards of dirt we have to recreate the valley, put a dollar value on it, and that's the cost of recreating the valley. Now what about that stream that meanders along the bottom and is maybe five thousand years old? Well, again, if you wait for the Mighty Hand of God to dig your stream, it won't cost you any capital but you will wait for a long time. So, if we apply modern technology we can dig a stream at so many dollars a cubic foot of water.

Finally one lady stood up and said, "Ah, but Mr. Yannacone, only God can make a tree." "Yes, that is very true—only God can make a tree, but isn't there a forest twelve miles away behind this mountain that nobody cares about? Let's move it at so many dollars a tree; we can transplant the forest."

There we have it—the total cost of replacing this particular valley with its supporting ecosystem, if the highway planners made a mistake and the next generation would like that valley back.

Now, what can all of you do—especially scientists? Insults to our environment by modern technology belong in a new category of corporate crime. It is time to treat this crime as one that utterly dwarfs into insignificance crime on the streets in terms of people injured and killed because of dangerous machinery, pollution, and unsafe products. A major group of corporate criminals are those industries responsible for environmental degradation. As Dr. Clancy Gordon said, "Those companies are pushing mankind to the brink of environmental doomsday."

There are four conventional appeals to the law for the protection of the environment: the first, and in theory the simplest, is through the legislatures of the several states and the Congress of the United States. If this approach is successful, there will be, of course, no need for other than occasional interpretive litigation. The ways of the legislature, however, are slow and ponderous. Many of our natural resource treasures are in immediate danger of serious, permanent, and irreparable damage.

The Florissant fossil beds represent a classic example of legislative ineffectiveness in a crisis situation. At stake were the unique and irreplaceable Florissant fossil beds, a 6,000-acre area 35 miles west of Colorado Springs, where seeds, leaves, insects, plants and fish from the Oligocene period 34 million years ago are remarkably preserved in paper-thin layers of shale. These fossils, studied by scientists from all over the world, are the richest of their kind anywhere on earth. More than 60,000 insect fossil specimens and 144 different plant species have already been found. The Florissant fossils are considered by many scientists to hold the key to determining the ultimate effects of air pollution on climate, since the air pollution from the volcanic activity that preserved the Florissant specimens was associated with a sharply cooling climate in Colorado.

Following a subcommittee hearing at Colorado Springs on May 29, 1969, the United States Senate unanimously passed a bill establishing the Florissant Fossil Beds National Monument. But while the Congress was deliberating, four land speculators purchased over half the land to be included within the national monument and announced that they intended to begin bulldozer excavation of roads to open the land for development immediately, unless the land was purchased by government or private groups.

The Defenders of Florissant, an ad hoc organization of scientists and citizens dedicated to protection of the fossil beds, finally turned to the courts, filing suit "on behalf of all the people of this generation and those generations yet unborn who might be entitled to the full benefit, use and enjoyment of that unique national natural resource treasure, the Florissant fossil beds." They sought a temporary restraining order prohibiting the disturbance of the fossil shales by speculators until such time as Congress had completed its deliberations.

On July 9, the United States District Court for Colorado held that no federal court could interfere with the absolute rights of private property ownership, and the Defenders of Florissant appealed to the United States Court of Appeals for the Tenth Circuit.

"This court must not countenance destruction of a 34 million year old record," the Defenders argued, "a record some would say, written by the hand of God." But the court questioned its own power to grant a temporary restraining order and demanded to know "what statute does this excavation violate?" The Defenders conceded that "there is no direct statutory protection for fossils," and the court immediately inquired, "so what right have we to control the use of private land unless there is a nuisance perpetrated by the owners?"

The Defenders' only remaining argument was to point to a fossil palm leaf that had been discovered at the Florissant site and plead: "The Florissant fossil beds are to geology, paleontology, paleobotany, palynology, and evolution what the Rosetta stone was to Egyptology. To let these irreplaceable fossil beds become the basements for the A-frame ghettos of the '70s is like wrapping fish with the Dead Sea Scrolls."

In a precedent-setting ruling, the court of appeals restrained the speculators from disturbing the fossil beds, but the temporary restraining order terminated on July 29, 1969, and on that day the district court heard testimony and argument for a preliminary injunction.

Meanwhile, during the intervening two weeks, Congress had cleared the bill through a subcommittee of the Committee of Interior and Insular Affairs and now the bill was pending before the entire committee prior to release to the House floor for action. Nevertheless the district court again held that there is nothing in the Constitution to prevent a landowner from making whatever use of his property he chooses.

Again it was necessary to appeal to the court of appeals, and at the hearing the speculators stated that they only intended to scrape off the top layer of the fossil shales and that would still leave as much as twenty feet of shales remaining. The Defenders argued, "You could just as well say scraping the paint off the Mona Lisa would cause no irreparable damage because there's still more canvas underneath," and again the 34-million-year-old fossils were rescued by a last-minute court order. A preliminary injunction was granted by the court of appeals as the bulldozers were poised at the boundary of the national monument.

Although Congress finally passed the bill, the difficulty with the legislative approach to environmental protection is best summed up in the words of the clerk of court of appeals, "Will you please get that bill through Congress soon and give us some rest."

Many legislatures, recognizing the delay inherent in the legislative process, attempted to meet the needs of our modern technological society by creating administrative agencies, to which they ceded some of the powers of the legislative, executive, and judicial branches of government in order to give speedy effect to the will of the people as manifest by act of Congress.

Unfortunately, the administrative approach carried within itself the seeds of its own abuse. Any administrative agency, no matter how well intentioned, is not a court, but is in fact its own judge, jury, and executioner—all in the public interest, of course. Its narrow jurisdiction and mission-oriented viewpoint make it inherently incapable of considering environmental matters with the requisite degree of ecological sophistication.

The *Scenic Hudson Preservation* case[354 F.2d 608 (2 Cir., 1965)] marked the fork in the road for those concerned with the protection of our environment and the legal defense of the biosphere. The Second Circuit Court of Appeals held that the Federal Power Commission should hear evidence on natural values in addition to the economics of electric power generation and distribution.

The tragedy of the *Scenic Hudson Preservation* case occurred when the Scenic Hudson Preservation Committee yielded to the Federal Power Commission jurisdiction of the natural resource aspects of the Consolidated Edison application, cloaking the FPC with a mantle of ecological competence it does not possess and cannot attain within the limits of its statutory mission. The conservationists in their all-consuming desire to avoid challenging established bureaucracy yielded to the Federal Power Commission the ultimate power to make ecological judgments binding on generations yet to come, judgments the FPC is constitutionally incapable of making.

Conventional tort litigation represents another avenue of appeal to the law on behalf of the environment, yet this avenue also leads inevitably to questions without answers.

Just what is a natural resource? Is it something that can be taken from the earth, then wasted, squandered, or used as the source of private fortune, or is it something that belongs to each of us as trustees for future generations, to be used wisely by whoever might hold nominal title at any particular time?

What do you do about a toxicant like 1,1,1-trichloro-2,2-bis (parachlorophenyl) ethane—DDT—which is ubiquitous, distributed throughout the lipid tissues of every living element of the biosphere? What do you do about a toxicant whose toxic effects cannot be demonstrated as the proximate cause of any particular personal injury or disease?

How do you balance the need for advancement of aviation, represented by the development of supersonic commercial transports, against the needs of the general population for privacy and freedom from the shock effects of sonic boom?

What do you do when a municipality decides that the highest and best use of the mighty Missouri River is an open sewer?

What do you do when the Army Corps of Engineers decides to drown the Grand Canyon or most of central Alaska, or insists upon destroying the delicate ecological balance of northern and central Florida with a canal that is no longer needed? I know of only one answer!

We must knock at the door of courthouses throughout the nation and seek equitable protection for our environment. Let each man and every corporation so use his own property as not to injure that of another, particularly so as not to injure that which is the common property of all the people, and let no wrong be without a remedy!

In 1966, a citizen sought equitable relief from a toxic insult to the community ecosystem, suing not just a local mosquito commission using DDT, but 1,1,1-trichloro-2,2-bis (parachlorophenyl) ethane—DDT itself. [*Yannacone* v. *Dennison, et al.,* 55 Misc.2d 468, 285 Supp. 2d. 53]

Finally in a New York court of equity the full weight of scientific evidence against DDT was presented to the social conscience of the community in a forum protected from the political, economic, and bureaucratic pressures that for twenty years had successfully suppressed that evidence of DDT's worldwide damage to the environment. At long last the agrichemical-political complex was forced to put its propaganda to the test in the crucible of cross-examination.

Three years later, at Madison, Wisconsin, Dr. Harry W. Hays, director of the Pesticides Registration Division of the U.S. Department of Agriculture, testified: "If the data appear to us . . . to be adequate . . . the product is registered. We look at the data, but we don't do it analytically. We don't check it by the laboratory method." At last Americans were told that the Department of Agriculture relies entirely on data furnished by pesticide manufacturers and does not conduct any tests on its own.

The incredible lack of concern for the safety of the American people became apparent on further cross-examination when Dr. Hays admitted that if a pesticide was checked at all, it was checked by an entomologist only for its effectiveness against the target insect and not for its effect on beneficial insects or fish and wildlife. "We don't assume that the intended use will cause any damage," he explained.

Moreover, Dr. Hays further admitted that although he has personal knowledge of scientific studies showing damage to fish and wildlife from DDT, USDA is "not doing anything" about possible environmental hazards from the pesticide. (Dr. Hays proudly stated, however, that the Department of Agriculture is completely responsible for the registration of pesticides and for determining whether they may be shipped in interstate commerce. He reluctantly admitted that the public has no access to USDA records of pesticide registration.)

Only in an adversary judicial proceeding was it finally demonstrated that the United States Department of Agriculture is really serving the agrichemical industry and not the American people.

At this time the environmental interests of civilization can be protected only by direct attack upon those actions which can cause serious, permanent, and irreparable damage to our natural resources.

Only by asserting the basic constitutional rights of all to a salubrious environment in courts of equity throughout the nation, can we marshall the weight of scientific evidence in defense of our environment.

In the struggle to protect natural resources against the predations of such short-sighted, limited-vision governmental agencies as the Corps of Engineers and the Department of Agriculture, any attack upon agency decisions must *not* be based on damage to a particular private economic interest.

The Everglades cannot be saved from the Army Engineers by showing the potential loss of income to hot-dog vendors in the Everglades National Park as the National Audubon Society attempted to do in the *C—111* case. Nor could the Florissant fossils have been saved by any unscientific appeal to aesthetic sensibilities.

Since the development of modern scientific methods of determmining the real social cost and actual social benefits of a "public improvement," it is obvious that the only way to save any national natural resource treasure is to establish with competent scientific evidence that the resource represents a unique and essential element of our environment and belongs to all the people not only of this generation but of those generations yet unborn, and that the actions of the defendants will indeed cause serious, permanent, and irreparable damage, in the classical equity sense, to this national natural resource treasure.

Conventional conservation education will not save the Everglades, the Grand Canyon, the Yukon, the Oklawaha, the Florissant fossil beds, or any other natural resource which has become the object of private greed or public blundering.

Only imaginative legal action on behalf of the general public, in class actions for declaratory judgments and injunctive relief, will lay the matter before the conscience of the community in a forum where the conflict can be resolved and the evidence tested in the crucible of cross-examination.

Although we in the trial bar consider it the major part of our professional obligation to avoid litigation and encourage settlement whenever possible, there is a time not to settle. There is a time not to compromise. There is a time to try the case.

Where the issue is of national significance and vested interests are arrayed against the common public right, trial is inevitable. Only during a trial before an impartial court or jury with skilled advocates presenting the evidence on both sides of the issue can the conscience of the community balance the equities and ultimately determine the matter on the merits.

The time has come for all of you, especially scientists, who would defend your environment, to assert your right to a salubrious environment as your own natural right and the right of all men, as one of the fundamental unenumerated rights guaranteed by the Ninth Amendment to the Constitution of the United States and protected by the due process and equal protection clauses of the

Fifth and Fourteenth Amendments to the Constitution. The time has come for all of you who would protect the environment to knock on the door of every courthouse in the land and establish once and for all that our national natural resource treasures are a public trust held by each generation for the benefit, use, and enjoyment of the next.

Experience has shown that litigation is the only non-violent civilized way to secure immediate consideration of the basic questions of human rights. Litigation seems to be the only rational way to focus the attention of our legislatures on basic problems of human existence—the only way, that is, short of bloody revolution. If we close the door to the courthouse, we open the door to the streets! And those of you who would make wise use of our natural resources, and protect the basic elements of our environment—air, water, and diverse viable populations of plants and animals—look to the history of the human rights struggle in the American courts.

Look to the success of the American labor movement and the surprising corporate survival of General Motors, in spite of the courts' recognition of the rights of the United Auto Workers.

Most of the major social changes which have made this country a finer place to live have had their basis in fundamental constitutional litigation. Somebody had to sue somebody before the legislature, in enlightened self-interest and recognizing the public benefit, of course, took action. Our adversary system of litigation as the method of presenting and testing evidence has been the touchstone of Anglo-American jurisprudence for centuries. It has always been the last hope of the citizens seeking redress of a public law. Now it is the last hope of the environment.

Why should scientists go to court and subject themselves to test by cross-examination from country-type lawyers? One simple reason. For nine hundred years the courts of England and America and before that of Rome have assumed that the absolute truth of a contested issue is unknowable at any given moment in time. They have assumed that every vested interest will tend to favor its own cause. So they developed a system that was immune to the self-seeking and self-serving of the advocate. And at the same time they sanctioned the position of advocate—unabashed, unashamed, tactless advocate. They put two of them on opposite sides of an issue and let them question, abuse, and even treat badly the representatives of the other side.

Where the advocates are generally equally skilled, finally at the end of a long session of inquiry, some general elements of truth will out. They will represent the best knowledge available at the time of trial. And for those of you scientists who are afraid to testify because the court will not understand you, I'll quote you in the words of the court a rule of law that is well over thirteen hundred years old and represents the basis for the evaluation of scientific or expert testimony. The question is asked: "Doctor, can you state your opinion with a reasonable degree of scientific certainty?" What does the court mean? What is a reasonable degree of scientific certainty?

"That degree of certainty called 'reasonable' from a scientist or a medical doctor need not soar into the icy stratosphere of absolute truth. It is enough, if earthbound and flat-footed, it merely tips the scales of more probable than not." Those are the words of a court summarizing thirteen hundred years of law on expert testimony. Now, ladies and gentlemen, the courtroom is the last arena where the individual citizen can meet mighty government or big business and hope to survive. The environment, the very breath of life needs the protection of the courts. As Woody Guthrie sang throughout this country in the thirties

This land is your land, this land is my land,
From California to the New York Island,
From the redwood forest to the gulf stream waters,
This land was made for you and me.

It does not belong to General Motors or any other soulless corporation, and it certainly doesn't belong to the Atomic Energy Commission, the U.S. Army Corps of Engineers, or the U.S. Department of Agriculture. The air we breath, the water we drink, and our natural resource treasures must have their day in court. Remember that Thomas á Becket and Thomas More are only two of those who died that the citizen and his environment may have their day in court.

Chairman:
As administrator of the Model Cities Program, Boston, with a professional background in engineering, Mr. Parks brings technological and deeply human insights and experience to the discussion of urban and suburban problems. His long involvement in the engineering and planning aspects of urban development, and especially the social and human facets of these never-ending complicated problems, have shown him to be a man of stature. Through the Model Cities Program, he is setting up comprehensive projects in all problem areas of the model neighborhood in an attempt to improve the quality of life in the neighborhood and to develop new methods of government for use throughout the city. This model neighborhood is an area of two thousand acres containing 63,000 residents in Roxbury, north Dorchester, and Jamaica Plain.

Urban Problems and Technology
Paul Parks

I feel so overwhelmed. In the last few weeks I have heard a lot about DDT and what it is going to do to us. I heard the head of Health and Hospitals for the city of Boston tell me that we are slowly dying because there is more and more of this and that in our tissues that may cause our complete demise. And I hear people talk about pollution and about the fact that the water we drink is polluted. I am worried because nobody really knows how to solve these problems. Having been involved in sanitary engineering some time ago, I think about all the methods we are using to try to stop pollution. I think about the smog problem and how it is doing away with us. I think about the people who smoke too much and what that is doing to them.

In spite of all these depressing problems, I am optimistic enough to believe we ought to be dreaming about a better future for people living in urban centers. I don't know whether our pollution or our dreams will win out, but I still somehow feel that we can in fact make the physical and emotional environment that people live in better than it is today. I believe that we can and must begin to plan ahead since it seems obvious that by and large most of our citizens will be living in urban centers. Many of us all along have felt that we were in a kind of safe catchment area living outside of the major city but peripheral to it—close enough to move into it and out of it with ease. Now, however, we are beginning to be captured by this great octopus called the city. It's reaching out and pulling us in. Suburbanites are being forced to make decisions. No longer are the suburban areas comfortable, nor can the people who live there be comfortable in feeling that they don't have a problem. Moreover, they must play a part in the solution of urban problems because they are now, in fact, urban.

Many people predict that by 1975, 80 percent of our population will be living in giant urban centers. We must envision those great cities that people talk about as probably running down the entire eastern coastline in a sort of unbroken chain. People must try to find a way to live in this massive physical environment.

I worry a bit when I hear people talk about the systems approach as the solution to our everyday problems. Somehow or other we get trapped by our own rhetoric when we say that systems analysis is the answer to solving people's problems. Let us now discuss these problems and consider the role scientists and engineers may play in their solution. In making use of the systems approach our methods must be alterable and flexible. Many times the scientific answer is not necessarily the human answer and has to be tailored in some fashion. And that is a very difficult thing to get across to those working in systems analysis. I know that my staff of about sixty systems analysts is constantly being frustrated because they can't come to grips with the fact that there are other things that they have to look at as constraints rather than the rules of their particular field. I don't trust a system to solve the problems of our urban cities. I do, however, trust the systems analysis to be a part of the total solution.

Flexibility must be the dominant characteristic of our system as it relates to the individual in our society who needs services. Such a person in need at present doesn't feel comfortable with our system. He doesn't feel he has access to it. It is too far out of reach for him. Many of us smile and say, "You know, we ought to train people to understand the system." Let me just say that for those of us who say that we have an understanding of how one may approach solutions, who are advocates for changing patterns, who are advocates for a better life for our urban people, it is incumbent upon us to articulate our position to people generally in such a way that they can understand it. Whenever society does not respond, we must realize that possibly it's because we have not explained our situation or our thought process in a way that enables others to understand.

We must develop our flexibility to think and feel differently about some of our welfare problems. Let me illustrate. Last night I was talking to a group of people who said, "We have a grand new housing project in our community. But this lady was admitted with her seven children, each by a different and undetermined father. We are outraged that she has been given an apartment in this new project. She does not represent the kind of person that ought to be living in this complex."

I asked them at this point, "What is the alternative?" As they spoke I heard them say that this woman had been living in a cold-water flat where the heat was spasmodic and she had all kinds of difficulty in dealing with her children. Somebody decided that they were going to have her live in decent, safe and sound, sanitary conditions for a change. But now people were saying something else to me. They were saying that the immorality of this woman ought to preclude her from sharing in anything of value in the system. The spokesman for the group was, lo and behold, a scientist. I am not saying that he was a typical scientist. But what distressed me was that here was a man eminently qualified in his field who was supposed, by profession, to probe things deeply before coming to a conclusion. The fact that he was an expert in this field of science lent a degree of credibility to his censure of this dear lady who had not conformed to the rules of the society within whose project she was living. Though he was not an expert in the field of social sciences, anything he said about people, or about anything, had a certain credibility and exerted an impact on the thinking of others. He certainly had developed little flexibility in his manner of thinking about social problems. Each of us has a great responsibility to develop this flexibility in our manner of thinking about how people fit the system.

But what are we going to do as this city spreads and continues the urban sprawl? How are we going to design the kind of environment that people will feel comfortable living in, where people will have access to their political structure? How can they manipulate their enviroment so they can vote out those people with whom they do not agree or who they believe are not serving their purposes and best interests? How are we going to establish that kind of city? We don't have it now.

I look at the political structure of our city and I see it as almost primitive, because it is inaccessible. This makes me think about the black people who began to scream "black power" a few years ago. A lot of people got all upset about this, as if it were a brand new concept. These people said that they want to be able to handle their own environment. They want to be able to make a decision about who is going to govern them. They want to be able to make a decision about their life quality and to be able to be heard by someone who can make a proper judgment of their needs. After listening to Stokely Carmichael and a lot of othor pooplo that I have a deep respect for, I remembered something from my early American history. I wasn't quite sure of the rhetoric or the language, but I read some of the early colonial rhetoric and it sounded so familiar. The colonists faced the same issues and used the same language that we have heard again in recent times: the irrelevance of the system; lack of access; the folks who make decisions are too far away; they are sitting across the water; they don't know about my problems; they are not down here where my problem is; and what gives them the right to make decisions about how I am going to live without talking to me first about it?

Today we have not been able to design the kind of political structure in the city that gives people the feeling of access. It is too far away from the users. The people are saying, "I don't want to deal with it." I say that we have got to think about a new way, a new process, a new direction in developing political structures in our city if we are going to make the city habitable for people. And it can't be so complicated that people don't feel comfortable with it. I am saying that we must look for a total reevaluation of whether we ought to continue with the city council, alderman kind of government.

We ought to be thinking about government for a city that stretches out from Boston to New York with millions of people housed in that city. We're already looking at a New York City that is totally irrelevant in terms of being a viable place for people to live in and somehow feel a part of. What kind of political system ought we to develop for a city of that magnitude? How do we structure it so that people will feel that *That Thing* which makes a judgment about them is close enough so that when it makes its decision, it does so in light of the needs of the people affected?

Now, I am not sure how to do that, but it is one of the questions that I present to you here. We have some ideas about decentralization of services. We are talking about different models. How do you set up a political structure to govern the cities so that people will feel that they have a part in it?

Let's look at one of our most important problems, education. The educational system was never designed to deal with the kinds of students that it is being asked to deal with today. And I am talking about students ranging from the elementary school child to the university student. We have a system that has been loping along for a long time that said that if you come from the right kind of family, that if your parents have had a prior educational experience, that if they fit into a kind of normal mode, then you too could fit into the system. In addition, tests have been devised to describe a person's ability to learn and to discern what a person's potential may be. The great failing in each one of those tests was the fact that we could not measure a person's drive to succeed, his drive to overcome even those things that we thought were insuperable obstacles. We are not able to separate what is an environmental force in determining a person's ability to achieve as opposed to what is native talent.

So we said that we could deal with certain kinds of people because they were on our standard-deviation curve at the right point and, therefore, we could somehow see them as successes. And everybody knows that the person who teaches in the classroom needs to have successes. If you teach her to look for those successes, somehow or another she can walk out of there and say that her period of time in there was worthwhile. But there is another important dimension to education and learning that we never dreamed of before television and instant worldwide news. Television brought events from every part of the world to our doorstep. Television began to bring Biafra close to us, began to bring Czechoslovakians throwing rocks at the Russian tanks close to us, began to bring the war in Viet Nam close to us. Though we can't smell and feel these events, we can certainly see them and get emotional about them.

In recent times, people have been aspiring who have never done so before and who never thought that the system would allow them to aspire. We now observe that a lot of people are beginning to ask for access to the school system. The school system wasn't ready for them so the government responded with things like Head Start, saying that somebody was culturally deprived. But many of us asked the question as to who evaluates a culture and how do you define culture? Is mine just as relevant as someone else's? After saying that, we began to realize that if you are going to live in a kind of a society that we have attempted to develop, there are certain things you must do if you intend to walk along with the system.

In spite of those who flaunt the system, most of us are locked into the widely accepted cultural values of society which may often be unrealistic. That's the meaning of the story of the teacher who says at the close of the school day, "Go home now, little children, and get momma to help you with your new math. Let her begin to tell you how to handle the null sets and all those kinds of things." And mother sits there aghast because she has even forgotten her long division, or maybe never had it. Here comes little Johnny with his homework because the system said that mother and daddy should know what is going on inside and be able to help their children so that when they come back to school the next day they will have had input of parents. This is merely another proof that the greatest pool of unskilled labor in America are American parents. We are expected to be all kinds of things and we just don't have those skills. We really aren't trained for the job. We really haven't had the ability to sit back and say, "These are the conditions under which one becomes a parent." We ought to look at marriage philosophically but, unfortunately, we haven't developed any research papers on how one relates to his spouse. When judged by those people who sit inside the school buildings, we come up very short.

The world has been so full of change in recent years that the one block which was home, school, and my whole world when I was a child is very different in today's changing times. The dimensions of everything have changed and we must learn to adjust to change. How do we redesign the school system so that it will be relevant to the children, so that it will help the parents to understand what the educational process is all about and be able to participate in making judgments about what their children ought to be learning? How do we design a public school system that somehow does not make early judgments about the ability of a child to learn and give him the full opportunity to grow? Ought we to have general high schools, or ought we to have special high schools? Ought we to have even the high school concept? What kind of school system must be devised for the massive city that I talk about that will stretch down the eastern seaboard? What kind of educational system ought we to be thinking about now that can survive in that kind of environment and be a viable structure in it?

The universities are being plagued with change now. People can say, "I don't like the hippies." You can talk about the fact that our young people are irresponsible, but the guys who threw the tea in the harbor were irresponsible, too. You can say of these students that they aren't really serious; they are playing games; they are doing this because this is the thing to do. The idea of protest; the idea of picking up the dean and carrying him out and sitting him in the streets in front of the administration building is the *in* thing. A lot of people are saying, "That's all the students are doing; it's like what we did when we used to swallow goldfish a long time ago."

If you believe that, you are making one of the greatest mistakes of your life, because there is a difference. What is the meaning of this change in attitude on the part of students? They are questioning our competency; they are questioning our morality. They are asking: "Why should I learn this? Why should I learn this in this manner? Aren't there other ways I can acquire this same bank of knowledge?" We say, "No, because there is a credible way that you ought to learn because that's the status of education."

They are saying, "So what about the status of education! I don't care about it. What really is important is what I learn, and I ought to have the ability to participate in making judgments about the process by which I learn." We are saying, "No, we are not ready for that." You know what happened to us—we used to have the well-disciplined student coming into the university, to whom the university could play a parent role. We don't have well-disciplined children by that dimension and definition anymore and so we decided we were going to deal with some children from the ghettos. We decided we were going to deal with some poor kids because we decided they ought to have access to education. Now, maybe we ought not to have decided to do that, but I think we should have. Once having made that decision, we have a different kind of student coming into the university, and this has an impact on the other students in the university.

A further question is: How do you establish a high-quality educational system that will give equal access to all of the citizens who may wish to participate in it? How do you design a system so that people feel comfortable with it and feel that they have a stake in it? I contend that if one doesn't have a stake in something one tends to destroy it. We must look at these dimensions and discover a method to produce an educational system for this megalopolis that will sprawl up and down the coastline.

Another important problem area is that of health. I can show you an area not far from here in which 80 percent of the people living there have never had a minimal physical examination. They are ill every day but have no place to take their problems. They know that if they go to the waiting rooms of the major hospitals they will sit there for hours because they don't know how to deal with the system. The system says that if you call on the phone you can get an appointment with somebody so you don't have to wait so long. But a lot of people don't know that that's one of the rules, so they go down to the hospital with their illness, and they sit there all day long waiting to be served. Then they build up a resentment for that system which is supposed to be caring for them but doesn't. They go back and tell their friends and neighbors that the place over there is bigoted and horrible. Then these people who hear such remarks, when they get sick, don't go because they don't want the embarrassment. Human debilitation develops in such a situation. They don't deal with the system; they just sit there sick and talk about how bad it is.

One of our problems today is the neighborhood school concept which may well have been our system's greatest mistake. However, an even greater mistake was not having developed the neighborhood health center. Many people don't go to work because they are not healthy. Many people don't stay on their jobs consistently because they are not healthy. When the mother first conceives, she doesn't have the right kind of diet; she doesn't have the right kind of information on how she ought to handle herself during this period to give her child the greatest opportunity to come into the world and be able to cope with this environment. There is no place for her to go. Oh yes, we have a few clinics. But there is no clinic for the average person, where all he must do is to talk through a door and have a service delivered to him. We don't know how to do that yet. We don't really know how to provide health services at a level where everybody feels that he has access to it and understands that it isn't a dangerous thing, that it isn't a harmful thing to participate in some sort of health care service. We aren't equipped to recognize when a child needs to have a doctor's attention or when a father needs it.

I can show you an area where there isn't one health center for an adult male. We have quite a number of child care centers—not enough, but we do have them. We can take care of well babies, but we don't really know how to take care of sick babies because we don't have a place for them to go. The design of our programs did not allow for that.

I am giving you this dimension because I think one of the things that we must address ourselves to is how we provide health care, notwithstanding AMA, notwithstanding anyone else. How do we provide health care that is available without charge to the individuals who live in a city? Until we come to that point, we are not going to be able to serve the people that we have to serve. One can talk about health insurance and one can talk about a lot of designs. The only thing I want in the design is that the individuals who use the health center or the health services do not have to pay for that service, because one ought not to have to pay for continued existence. One ought not to have to pay for his life! I think that is what we are really talking about.

I am not concerned whether you think I am talking about socialized medicine or some other system of care. I am talking about the appalling fact that far too many people are ill who can't be cared for, who just get bogged down in the middle of the night because they have no place to go to get help. There isn't a doctor within miles. Every structure is a hostile one for them. At the same time we hear people say to them, "Get up by your bootstraps— you ought to be able to make it." But I tell you that everything is designed to keep them from making it!

I suggest we look at the structure of health delivery. How do we develop a system that is going to deliver health service to all of the people without their having to pay for it? One may say to me, "I think that your welfare system is a horrible dole system." I hear people say that those people ought to get up and go to work when all of us know that 85 percent of the people on welfare couldn't get up and go to work tomorrow if they had to.

How do we deal with the problem of those people in our system who don't conform to our system? What about that mother who sits around home doing nothing more than having several boyfriends, and ends up pregnant, maybe pregnant by all of them over a period of time. In our society she is regarded as an evil human being. Our whole moral concept gets bogged down around this whole area because when most of us say morality, we think about sexuality. But perhaps more important, there are other kinds of immoral things that are done to people in the city that have nothing to do with cohabitation. We can't tell that mother that she must not have that next child if she can't come into a clinic and have her problem taken care of. We are going to have to look at the abortion laws. We're going to have to look at the birth control laws because we can't criticize at both ends. We can't call this woman immoral unless we make the decision that any woman who has over three children ought to be put in a gas chamber somewhere so she won't procreate anymore and cause us any more of these problems.

A man said to me the other day, "Why should I pay taxes to support a slut like her and her children when she won't work?" I told him, "If nothing else, just remember when there are more of 'them' as you see it, than there are of 'you,' then you are in trouble." At that point the system may well be destroyed. If you don't think of it in any other way, think of it from the point of view of your own self-protection. I hope that is not what you come to, but if that is the only way you can see it, see it that way. You can't tell people that they should restrain themselves, that they should maintain a kind of morality accepted by the rest of us, when there is no payoff at the end of the road. Nothing in that girl's life is going to be better if she lives it in any other way. There is nothing that is going to change the life quality of such individuals, so why the devil should they play in your ball game?

Somehow we must define the ball game differently. Somehow we must give people access. I think about people who feel they ought to be able to walk in a door that leads somewhere. A woman walked into my office the other day and said that she was a month pregnant and she would like to find some place to go to get rid of it because she wasn't married. She didn't feel that she would be able to deal with this child. Now, the clergy and others may well argue against such solutions; and I am not suggesting how we are to solve such problems. I am just trying to say that there is a problem and that somehow we ought to find a way to give to a woman such as this access to a solution to her problem.

Let's carry our analysis one step further. When she has that child, all the rest of us back away. None of us has responsibility at that point. We can all tell her what she can't do and what she can't participate in, but once she has the child she is on her own. She must fend for herself because the system doesn't provide anything for her then except the welfare system, and a lot of people are angered at supporting a welfare system, especially Aid For Dependent Children (AFDC). At the end of the road, the only payoff she is going to get is public censure. She is condemned by the system to a life outside the system—the same system which refused to give her relief for her problem.

Criticizing those who somehow get caught up in such problems is not going to change the situation. Nor will legislation change it. We must learn to solve such problems in other more effective ways. A fact that we must face is that there will inevitably be people who will never be able to cope with their environment. Some will find it so complicated that they can't understand it. Somehow we must find a solution to the problems of people for whom the system, as we know it today, is too complicated.

To achieve this goal we have to think about a national minimum income of some sort. How can we solve the problem of providing a base income for the people of this nation who somehow or another can't keep pace with the system at any given time? There are people who become elderly in the system who have played a major role in the system and ought to be able to grow old in dignity. How do we design a system to care for this group of people? How do we design an effective system of police protection for the massive cities?

Today I have asked a lot of questions, but I think the questions make sense. These are the questions that challenge those of us who have chosen to deal with the changing urban environment.

We must devise a system soon by which people in the sprawling coastal urban complex can coexist with each other in a decent physical environment. Today I have not talked about the race issue because it is a by-product of the frustrations of people trying to exist in a very complicated system that makes judgments about people as they come to the table to ask for their share and to lay their stake.

If we can solve those frustrations before DDT, smog, and polluted water destroy us, perhaps we may then find a way for people to begin to deal with those other problems. Only then can the city of tomorrow be a viable place where people will feel that they have a stake in it and a right to exist in it.

Chairman:
The pattern of his life as a productive research chemist has several times been interrupted by the call to various posts of academic, governmental, and industrial administration. After B.S. and Ph.D. degrees in chemistry from Harvard University, Donald Hornig was early recognized at Brown University and Princeton for his diversified talents of outstanding research in theoretical chemistry, shock and detonation waves, spectra of cyrstals, and molecular spectroscopy, and for outstanding administrative ability.

Under John F. Kennedy he became a member of the president's Science Advisory Commission (1960-1968). He served as special assistant in science and technology to the president of the United States and as director of the office of Science and Technology for five years under Lyndon B. Johnson. In 1968, he became vice-president of Eastman Kodak Company, and was elected president of Brown University in 1970. This broad experience qualifies Dr. Hornig in a unique manner to discuss the topic "The Government and Science."

The Government and Science—Troubled Waters
Donald F. Hornig

This is a panel on science and the problems of society. Most of the discussion will probably be devoted to the acknowledged current problems—pollution of the environment, the population explosion, the problems of our cities and urban clusters, the transportation problem, and so on—and to the potential role of scientists and of research in meeting them.

These are important matters but I will leave them to other speakers. I want to address the problem of science itself. By science I mean organized knowledge and understanding of the physical, biological, and social worlds. Closely allied to it is technology, by which I mean our body of skills, mechanical and intellectual, our know-how. It is on the basis of what he knew and what he was able to do that man reached his present state. Now it is fashionable to make technology the whipping boy for our present problems—and this is just as appropriate as blaming medical science and modern agriculture for the worst social problem of all, an overpopulated globe in which growth in numbers is out of control.

We cannot backtrack in history. We cannot abandon billions to starvation, disease, and poverty. We need to know more and to enlarge our body of skills in order to keep this planet habitable.

The title of my remarks might better have been "The Only Alternative to Knowledge Is Ignorance." A civilized society depends for its advance on its knowledge and its skills—none in history has ever successfully proceeded from ignorance. So my thesis is that one of the urgent social problems now facing us is the maintenance of a strong scientific base for our society—otherwise we shall pursue will-o'-the-wisps based on good intentions and ignorance rather than develop plans based on knowledge and understanding.

This may be obvious to this audience but it is not obvious to the public and to our lawmakers. As a result, we are facing a deep-seated crisis in the support of science in our universities. You all know the superficial symptoms—the NIH budget has been cut back 10 percent, the NSF budget was cut back last year from $500 million to $420 million and this year's appropriation is $441 million. Cuts have also been felt by other agencies, such as Agriculture, AEC, and NASA. Most jolting of all, though, is the Mansfield amendment to the military procurement authorization bill which states that the Department of Defense may not fund "any research project or study" unless it "has a direct and apparent relation to a specific military function." Since the Department of Defense funds about 25 percent of the research carried out in American universities, and most of that support requires a broad interpretation of the amendment, there is a threat of imminent disaster to a lot of very high quality research.

I do not wish to argue here with Senator Mansfield, who is a very distinguished and able senator, although I am sure that if narrowly interpreted the amendment would be dangerous to our future security. Neither the nuclear studies which eventually made nuclear weapons possible nor the development of electromagnetic theory which made radar possible would have been funded under the Mansfield amendment. Still, one wonders why basic science should depend on the Department of Defense for its support. The justification for the general support of science by the Department of Defense is that, as became clearly apparent in World War II, the ability of the country to defend itself in the future will depend on its total scientific strength, just as it has depended for a century on its industrial strength. Given that fact, it is then the responsibility of the Department of Defense, if it is not being done elsewhere in the government, to insure that science is healthy and that our knowledge base is adequate. That was the original reason why the navy funded basic research in the 1940s.

Nonetheless, I felt then and do now that this residual responsibility of DOD needed to be taken over by civilian agencies in the long run. And indeed it has, as NIH, AEC, and finally the National Science Foundation have grown up. Defense support of university research has dropped from about 90 percent in the late 1940s to less than a quarter now. The Mansfield amendment would largely complete that process—and if it were accompanied by funds to the NSF to make the transfer possible it would be a step forward. The trouble is that funds are not being given to NSF and as far as I can tell will not be. The quality projects dropped by the Department of Defense will not be taken up elsewhere because they cannot be, in the absence of funds.

Funding is only a symptom of more basic troubles. Until recently science was funded from university endowments, by foundations, and to some extent by industry, all of whom were well acquainted with the subtlety of building our fund of basic knowledge and skills. Now the support of basic science has become the responsibility of the federal government, which supplies over 80 percent of the funds. The amount of support and the purposes for which science is supported have become political problems. And for that reason what was an internal matter for the scientific community has become a public matter, subject to public debate and decision.

What is happening is that under budgetary pressure the assumptions on which public policy has been based and the government organization which carried it out are now in question. As the programs were built up, it was assumed that each program, be it health, atomic energy, or defense, needed a strong, broad knowledge base and provisions for the training of students. That this was sound seemed self-evident, and the pluralistic system of support for scientific research and education which grew out of it was the basis for the most flourishing scientific enterprise in the world. Now that assumption is being challenged and the organization of the government to support science threatens to become unworkable.

On the one hand the utility of scientific research is questioned unless it is narrowly focused on particular targets and unless practical results are foreseeable in a short time. For example, I am told that the narrowing definition of health research is leaving important research on photosynthesis unsupported.

But the stupidity of a narrowly focused effort is obvious. What all-wise director of NIH would have foreseen that the solution of modern biomedical problems, such as DNA and RNA, would require the discovery of x rays and the development of x-ray crystallography for three decades before it could deal with the structures of biologically important molecules? Or would he have forseen that without developing the very large modern computer that would still be impossible? Would he have supported the physicochemical research which led to the variety of chromatographic separation and purification methods which have made it possible to isolate pure biological substances? The fact is that with a narrow definition of "health related" he would have done none of these things—and yet we know the progress of modern medicine has depended on advances in practically every field of science. The Mansfield amendment is similarly shortsighted in what will be germane to defense in a decade or two.

On the other hand, some of those who accept that scientific research is a "good thing" want it in every state and town, like a post office, and confuse it with rivers and harbors, viewing it as a kind of largess that the federal government distributes. In its extreme form this view says anyone who wants to do research should be entitled to public support. Too much of this in the recent past has probably contributed to loss of public confidence. We all know that a lot of "busy work" has been supported in the name of basic science.

The point is that the development of science must be guided by a sense of significance and relevance. Relevance to urgent practical problems is the appropriate criterion for applied science. Relevance to the training of students to carry on in the future is pertinent to all science. But basic research also has standards of significance and relevance to the advance of the whole body of knowledge which we need to achieve if we are to get our money's worth. Some research uncovers a few more facts which have little further significance; but important research opens the way to whole new areas of understanding. That is what we mean by relevant and significant basic research.

I believe strongly that what we need now is to recognize that naturally we must put trained scientists to work on urban problems, on pollution, and on providing better health for our people.

Naturally, we must encourage more scientists and students to become interested in these problems—or more important give the large number who are interested in them an opportunity to go to work. But in addition we must recognize the importance of science per se—by which I mean science which will most significantly increase our knowledge and understanding.

Any technological industry supports a considerable amount of basic research in its laboratories because it knows that it is always short of knowledge and that its future depends on its knowing more and having a better "bag of tools." And industry is hard-headed and cost conscious. For the same reason the country has to enlarge its foundations of knowledge and to add new tools to its kit. Until now we have been world leaders, but whether this will continue to be true I just don't know. We seem now to be backing away from the future.

My plea then is that we support scientific research in all of our problem-solving federal agencies and encourage each to be at least as broad in its perspective as a well-managed industry, but that in addition we set out to strengthen the effort to build our knowledge and understanding on a broad front as prudent insurance for the future. At a minimum this means a larger scope for the National Science Foundation, but if, as I suspect, things will get worse before they get better, it may require more sweeping reorganization of the entire federal science structure.

My thesis then is twofold: first, the society cannot afford to have its scientific research effort dismantled and disorganized and that is what is happening right now. Second, we cannot afford to say that all we need is more money. If science is to help secure the future of mankind, all of us, together with our public leadership in Washington, will have to learn to do things better as well as to urge constant growth.

What does that mean? When increasingly large-scale social problems have to be attacked (not solved) by debate, hunch, and hope, it seems plain that social-problem-oriented applied research has to be expanded. HUD has made a start, but where is the urban equivalent of the broadly based defense research program? We have all been alerted to our environmental problems, to the ecological problems, which may threaten our very existence one day, but does the Department of the Interior have a broad program to deal with these? Except for the U.S. Geological Survey, I haven't been able to find it. We have to put more, not fewer, people to work on urban problems, on pollution, on providing better health for our people. We must encourage gifted scientists and students to become interested in these problems—although it is probably more important to give the large number who are interested an opportunity to go to work.

But, in addition, we must recognize that all of these problems will be with us for a long time to come. They have no quick solutions, but with well-directed efforts progress is possible. And the key contributions to the most difficult problems will come from unexpected scientific disciplines, just as they did in the case of DNA, which I mentioned earlier.

Therefore, I repeat, if we are not to proceed from ignorance, we need knowledge to guide us, and this makes the healthy progress of science itself a major social problem if the future we might have is to be realized.

Discussion Session

Question (Professor John Maguire, Boston College): Toward the end of your presentation, you discussed the restructuring of science departments in a more meaningful fashion such as including classical physics and geophysics in one department. In this way you indicated that we might promote a more meaningful interdisciplinary approach to some of the problems that beset us. Actually many of the problems that beset society today are all-encompassing in their interdisciplinary nature. If formidable social problems such as transportation, for example, are to be resolved, resources of law schools, economics departments, physical science departments, social sciences and others must be utilized and be brought to bear on solutions.

I agree with you that the departments are all too compartmentalized, but how can we go beyond this narrow restructuring that you propose so that the economics department, for example, can interact better with the physics department? How do you propose that there may be a really meaningful multidisciplinary interaction at the university level?

Professor Wilson: Well I think one must clearly differentiate two aspects of the problem: one is to ensure students a broad choice of courses which they can take from different departments, and the other is to tackle departmental problems. I think the important thing is to shake up the departments so they think in new and different terms, although I don't think that it is feasible to completely reorganize the structure of the university. One can set up graduate and research institutes to deal with problems of transportation, but I don't think that you can go so very far in this direction, at least as regards undergraduate teaching.

My thought is that without changing a single lecture course in a university, it would still perhaps be useful to mix, for example, some chemists, physicists, and some biologists just to get people talking with each other along different lines. This would help them to guide students. At the moment, if a student wants to take chemistry, any chemist can tell him all the courses that are arranged in the chemistry department which can be taken in sequence and the student can learn a lot about chemistry. However, if he wants to learn about the interdisciplinary aspect of chemistry or the effects of pollution and its economics, the chemist may not have the least idea of how to direct him. So, I think it would be a good idea to have some slight changing of the departmental structures so as to broaden scientists' knowledge in related disciplines and consequently to help them think along broader lines.

Question (Robert Leefert, Michigan State University): Dr. Wilson, I have just started teaching in a large university of 42,000 students. We have a variety of faculty of many different interests. It would be very rewarding for me and for other faculty, especially the younger ones, to cooperate in multidisciplinary programs such as those referred to. However, it seems to be quite difficult to break through departmental lines right now, although there is mounting interest in such cooperative efforts. It seems that budgets are set up in such a way that interdepartmental programs and joint appointments are fraught with administrative problems. The student also has a problem in crossing department lines. I regard it as very important for a student in engineering, metallurgical engineering, and material science, which I teach, being a mineralogist, to get a strong background in physical chemistry. Yet, when such a student goes to the chemistry department, he must compete with chemists. The chemists treat all students like chemists, not recognizing that they are not chemists. They just want to learn an appreciable amount of chemistry to apply to their specific discipline. What would be your suggestion as to grading, for instance, to ensure that students will not be so afraid of flunking a course that they won't take it? This is the basic problem. I think students are afraid to take courses in other departments because the faculty specialists burden them unduly if they step out of their own narrow discipline.

Professor Wilson: I can see there are many problems in changing a university of 42,000 people. However, at Erindale, a new campus in the suburbs twenty miles from the main campus of the University of Toronto, where we have just under 1,000 students at the moment, I have recommended such changes as may bring these new structures into being, in the hope that it may set an example to other universities. I can see the problems, however, in changing a large university.

Robert Leefert: We have a Center for Environmental Quality being set up which will call on people in various departments to cooperate in research. This does not really solve the problem of the student or provide important answers to the question of grades. It is important to maintain certain standards, but I think that when a fellow is trying to stretch his brain a little bit in a different area outside a department, there should be some educational perspective developed so that a professor may be a little easier on him.

Father Skehan: I would like to speak to the point that was just raised regarding interdisciplinary programs and the manner in which one of our visiting professors, Dr. Cyril Ponnamperuma, has attempted to solve the problem of grades in an interdisciplinary course. In the spring 1970 semester, The Boston College Environmental Center through several faculty and lecturers from a variety of disciplines will offer a course, Chemical Evolution and Origin of Life. Students from the departments of Biology, Chemistry, Geology, Geophysics, and Philosophy will be enrolled. The grade for the course will not be by the conventional final examination, which would obviously favor a group from one science over another, but by means of a research paper, which would explore an area of particular interest to the individual student.

Comment (Fred Quelle, Office of Naval Research, Boston): I think one of the more powerful means we have of encouraging scientists to move in important directions is through financial support of research. Interdisciplinary materials research laboratories have been set up where physicists, chemists, metallurgists, and others are brought together to work on problems jointly. In general, I think that we have not done enough of this in the past, but this certainly is one means by which we can bring diverse groups of people together. Unfortunately, however, I think we have not yet set up mechanisms for providing adequate support.

Professor Wilson: I certainly agree that much is being done already, but I would merely say that more needs to be done, particularly in fields that are presently neglected and which are very difficult, perhaps not always attractive to scientists, such as the question of the earth's heat budget. The generation of so much power to meet human needs is producing heat and carbon dioxide and this acts as a shield in the atmosphere. The content of carbon dioxide in the atmosphere has gone up. Such problems are more complex than materials research or oceanographic research and the like, but they need investigation and need it very badly—research in many areas has been neglected.

Question (unidentified speaker): There is an apparent revulsion on the part of scientists to this adversary-type proceeding. Presently, Ernest Sternglass of the University of Pittsburgh has made a controversial statement about the damaging effects of nuclear testing and other similar statements. Because he's controversial, *Science* magazine has refused him publication; has refused to take his findings and put them before the scientific community. Can you comment on how we are to change this manner of proceeding or this attitude within the scientific community?

Mr. Yannacone: Sternglass testified in an action that is now pending against the Atomic Energy Commission before the United States District Court in Denver seeking to restrain the release of radionuclides from Project *Rulison,* the underground gas stimulation experiment. Dr. Sternglass and his allegations did not survive cross-examination in that court. However, the scientific community, with the so-called refereeing that goes on in many journals of which *Science* is one of the better, nevertheless fails to investigate adequately the basis of a claim, or to test adequately the hypotheses presented in papers. To this day, there has yet to be any real scrutiny of the pesticide question, particularly with reference to DDT, even approaching the magnitude, scope, or depth of the inquiry that was conducted in Wisconsin as an adversary litigation proceeding. The great failure in testing scientific journal articles today is that cross-examination, or as the scientific community prefers to call it, review, is not conducted with any reference to the nine hundred or a thousand years of experience the legal profession has had with the search for truth by dialectic inquiry.

There are three words that *we the lawyers* consider in testing evidence: Is it *relevant,* is it *material,* and is the witness *competent?* There are tests hundreds of years old that the evidence presented must survive before it can be accepted and relied upon in an adversary judicial proceeding.

While I think the Sternglass paper should have been published in *Science* so that the whole world could consider it, I do think that the hour and a half of Sternglass's cross-examination by the attorneys for the Atomic Energy Commission is more important than his paper, and more important than that, are the many hours of cross-examination of the Atomic Energy Commission experts who later disputed Sternglass's work.

Any scientist who knows his business, is telling the truth, and is willing to stay within the confines of his own discipline, although that means not being as narrow-minded as some people would like, need have no fear of the courtroom.

The most abusive treatment of scientists in the courtroom I have ever seen occurred with Dr. Wurster and Dr. Risebrough during the early days of the Wisconsin DDT trial. The industry cross-examination was conducted by attorney Louis McLean, the same man who wrote in *Bioscience*—another journal that lacks something in its refereeing process—that all pesticide opponents were either sexual perverts or lawyers seeking publicity. Nevertheless, Wurster and Risebrough were challenged for a total of six days of cross-examination. This cross-examination was ruthless, embarrassing, cruel—it was the worst that the legal profession can demonstrate. Yet, when it was all over, nothing had happened to either the scientific reputations of Dr. Wurster or Dr. Risebrough or the value of the evidence that they had presented on direct examination. In fact, their testimony was later supported by experts from Shell Chemical Company. Again, the court exists to protect the professional dignity of its witnesses, especially when them come from other disciplines. I have yet to see a scientist too badly treated, although there still are a great many scientists whose information is so weak it will not stand the scrutiny of even a poor country lawyer like me.

Question (George Schindler): Aside from my association with *Science,* I am also a member of the Sierra Club. Almost daily I have contact with the Corps of Engineers, highway builders, and the like. Frequently, questions arise in which we attempt to get legal help on the matters we are involved in. I think I see stirrings within the legal profession to try to do the sort of thing that you have been doing so effectively. The problem comes in perhaps the next few years, perhaps five or ten years. Can you suggest anything that will help stimulate those in the legal and scientific professions to be a little more forthright in offering their help?

Mr. Yannacone: The legal profession has been, to a large extent, abused by the academic science community as a cross between ambulance-chasing shysters and dull-witted corporate attorneys who don't know what they are doing. Fortunately, there is a new breed of lawyer entering the law schools and now graduating with background in some of the natural or physical sciences. Many of these young graduates are looking for futures in environmental law. The American Trial Lawyers Association, which represents the 33,000 working trial lawyers in the country, at their annual meeting last summer pledged support to the environmental community as such.

The way we are going to do that is with two programs. One is an environmental advocacy seminar program which we will take around the country pretty much at our own expense unless we can find some friendly foundation, a prospect which doesn't seem too likely at this time. We will take these environmental seminars around and expose our 33,000 members over a period of five years to the mechanics, the nuts and bolts, of how to handle local enviromental problems with some skill and with some possibility of success. The second program utilizes the systems approach, which we are going to consider this afternoon at the Society for General Systems Research meeting, and is drawing law and systems analysis together in an effort to integrate much of the work of the environmental science community with the legal process and the free enterprise system.

The big problem today is that most scientists have failed to relate their individual discipline and activities to environmental problems as a whole. Of course, most lawyers fail even to recognize that there are environmental problems until a particular client is injured. The development of a meeting ground can be catalyzed in large measure by magazines like *Science.* There are stirrings that law and science will be drawn together in the near future and, indeed, have to come together over such issues of environmental concern as the sonic boom, the Big Cypress jetport, highway route selection methods, the Atomic Energy Commission and its Federal Radiation Council Standards, and the pesticide problem. The difficulty is that some major scientific organizations and agencies fail to recognize the problem and ease the union of science and the law. I notice, for instance, that the AAAS recognizes a great many of the behavioral science disciplines as such. Yet it does not really recognize the law, while the law is older as a learned profession than all the other disciplines represented in AAAS.

Question (Maria Bade, Boston College): For a scientist, law is terra incognita or worse; we are all, I think, a little bit frightened of it. I want to mention three possible cases that agitate me as a biologist and invite your comment as to what might be done about it. The cases, or I should perhaps say problems, are in different stages of development, and many of them would appear to be outside the competence of local courts. The first one is Big Thicket in Texas. There you have an area which is beautiful; nobody has noticed that it needs conserving, but there are people who do want to exploit it. So, in order to spike the guns of anybody who might say that this is something that might have to be preserved, the exploiters have moved in first, with massive pesticide spraying to kill off heron colonies and chopping down the magnolia trees and so on. Later, they can say: "Oh look! This is a mess! But we are going to develop it. We are going to use it properly. You want to preserve it? What on earth for?" In other words, these things can happen before we notice. If we have to bring a case for preservation of every leaf, we are going to be in a lot of trouble.

The second problem is in Brazil, where for a great many years there was, and is, a law that was considered a model law for the protection of the wild Indians in the interior of the Amazon basin. The law said quite plainly that as long as a member of a tribe was alive to enjoy the territory claimed by the tribe, this territory could not in any way be used by anyone else without the consent of the people there. So what has been happening lately is that people have moved in with airplanes and started to kill off the tribes, and whole tribes have simply been wiped out with everything from bullets to machetes. Then they say: "Oh well, there is no one there any longer so now we can drill for oil, chop down the trees, or do anything else we like. We are all set."

The third problem is the case of the whales. There is an international agreement not to overhunt, but they are being over-hunted anyway. It now seems that the point of no return may have been passed with a number of the great whales; there simply are not enough left to meet and reproduce. Yet in another generation it may become clear to people that whales might have been used a great deal better than by hunting them to death—for example, they might have been trained to produce milk on demand. So here you have a case where an international agreement exists but it cannot be enforced.

Mr. Yannacone: There are American corporations involved in all of those problems. They are amenable to the jurisdiction of American courts. If you as biologists can show serious, permanent, and irreparable damage to the human or natural ecosystems involved, then the courts and the law can help you. I say to you what I have said to scientists before and more specifically to a group of scientists prior to litigation that I have been involved in: "Don't just sit there and bitch about the problems, sue somebody."

Question (Curtis Williams, Rockefeller University): In your perspective of the scope of the problems of the urban megalopolis, such as Florida, I wonder if there is a place for the small-scale project that may attack the specific problem. I recognize that this can't be a solution to the overall problem, of course. This question was stimulated by your discussion of systems analysis and by the fact that people need success to stay interested in what they are doing. You are asking for the kinds of changes in which there are going to be very few successes for a long time. How do you keep your constituency together during this long period of reevaluation? How do you find solutions to monumental problems which are not welfare problems, which are not health problems per se? These problems are fundamental to the makeup of society, ingrained and institutionalized. What is the place for the small project, for example, that would go into the community and identify the sources of lead poisoning? What is the role of the law community in dealing with a focal problem that may synthesize but really may solve only special aspects of the problem?

Mr. Parks: Our systems people and other planners periodically get frustrated because the payoff is a long way away. We start them on certain small projects, such as a health center that has to be opened. Whether that health center solves the problem is irrelevant. There are some people who need help, so we have opened several health centers. Thus, the systems people get involved with everyday problems. They will come in to work for us for a while and leave to work elsewhere—it's the level of frustration! The payoff isn't there immediately! They do a service for us when they are there; they are in tune, they know where *they* are going, but can't visualize where *we* are attempting to go. Somehow it isn't quite as gratifying as I would like to have it. I think that is a way of life. I think that is the kind of world we are going to live in for a while. Our agency is looking at tomorrow, looking at a redefinition of a long-range system. There are also urgent problems in the city. Thus, people can pick and choose whether to get involved in problems with long- or short-range solutions.

Two interesting things are happening in America, one of which is taking place in Atlanta. Atlanta would easily have had a black mayor, because the black population is growing in the city relative to the white population. But the Georgia legislature developed a whole-county government along the lines of metropolitanizing city government. We are getting trapped between these two chairs.

There is a valid need for metropolitanizing in order to diffuse the tax base and for other reasons. At the same time it is hurting the black folk, because, as we begin to rise in power in the city merely by reason of numbers, the balance is threatened by the move to metropolitan government. People who were opposed to metropolitan government for many years began to favor it when they saw that there would probably be a black mayor of the city. They began to advocate metropolitanization. This is one of our great frustrations.

Now you ask me about electing black folk. We should make a judgment about those who make the most sense. The greatest pool of talent left in our cities is in black communities because the talent of the white community, by and large, has disappeared to the suburbs. Talented blacks, then, have emerged as candidates. We ought not let their blackness stand in the way of voting them into office. That has happened far too often. At such a point, blackness becomes more important than qualifications. When this happens, everyone is cheated.

Mr. Yannacone: At the risk of insulting some of you scientists, there is a message hidden in Mr. Park's statement that he wasn't tactless enough to broadcast. I am going to tell you even at the risk of being tactless. What are all of you in the scientific community going to do when black people start reading the literature and find out that black people carry two and one-half times the amount of DDT as white people? What are you going to do when black people discover by reading the literature that most of their people are the ones that are dying of lead poisoning or their children are suffering mental retardation in our cities? What are you going to do when someone takes a look at the Philadelphia studies and finds out that the correlation between social disease, social disorganization, and air pollution is quite high and concentrates in the black community? What are you going to do when suddenly it is discovered by the black community and the people who live in the slums that it is the summation of a myriad of little toxic insults that makes the slums as unbearable as they are? What are you going to do when they discover that the Federal Radiation Council, the Department of Agriculture, and the Food and

Drug Administration admit that many of the environmental toxicants present today are damaging but then say in the same
breath that they are damaging only to a statistically small number
of the great American population? What are you going to do when
the black people and the people who live in our slums and our
ghettos discover suddenly that they are the statistically small
number that is being damaged?

When ten thousand people died as a result of the air pollution in-
version over Thanksgiving weekend in New York City in 1966,
the majority of them came from Harlem, East Harlem, and Bedford-
Stuyvesant. The majority were children and old folks. What are
you going to do when that kind of data is finally released by the
U.S. Public Health Service and appears on page 3 of the *Daily
News?* The time has come for those of you in the scientific community not to dabble in politics, where you are out of your element and basically incompetent, but to go straight back to the
laboratory and to science where you belong and start using your
scientific knowledge to solve these immediate environmental problems.

There is no legal justification for imposing such toxic environmental stresses on these people; to fly a supersonic transport
and stress these people even further with noise so that one-tenth
of 1 percent of the American people, the richest people in the
country, can fly a bit faster when the majority of the people can't
even get from work to their home with any ease. When is this going to stop?

You scientists have a duty to society. The only thing the legal profession asks of you is that you come forward, put your scientific
reputations where your conscience must be and where your heart
should be. Say what has to be said! Challenge the callous disregard of a heartless bureaucracy for this statistically small segment
of the American people. Demand that any industrial operation,
any governmental operation, any operation that has environmental

significance be conducted in accordance with the highest and best use of the national natural resources that belong to all the people. In particular, protect the people that are already stressed almost beyond endurance because of packing in our great coastal megalopolis.

Otherwise, you who sit in your university laboratories, on your green-grassed, tree-lined, birded campuses, from which many of you run away for two or three months of the year, as well-fed wards of the state may, when you come back, find other people sitting on that campus escaping from the city.

And when you try to continue your work, you will suddenly discover that the big question the law has asked others for twelve hundred years is being asked of you now—relevance. What is the relevance of your work? What is the relevance of the product of the money that goes into your program? What is the relevance of your research to the real problems of the real world? And as I am going to tell the systems analysts this afternoon, there is nothing wrong with systems analysis that compelling systems analysts to deal with real-world problems and come up with real solutions wouldn't cure.

Part II
The Scientist and Society
Chairman: Robert L. Carovillano

Chairman:
Franklin A. Long has a distinguished scientific record as a chemist. At Cornell University he is a professor of chemistry and the Henry Luce Professor of Science and Society. Formerly a vice president, Dr. Long is now director of Cornell's Program on Science, Technology and Society, a post he accepted in place of the directorship of the National Science Foundation. This program at Cornell was conceived as an organized response to matters of grave national concern. It focuses on problems of national and worldwide scope such as the relationship of science, technology, and public policy, including defense policies, world food supplies, ecology, population growth, and increased urbanization. In detail, the program consists of seminars for faculty and students, the development of new courses, one of which is presently being given by Professor Long, new curricula, guest lectures, and research.

Rethinking Scientific Objectives
Franklin A. Long

As the lead-off speaker for this program, I should probably try to identify the problems that will concern us during the rest of the discussion. It seems to me that I can do this by posing two short questions. The first question is: What is science for? The second question is: What are scientists for?

A more conventional approach might have been to ask the single question: What are the goals and purposes of science? But I did not. I ask two questions rather than one to emphasize that the goals and purposes of society for science and those of scientists for science need not be identical and often are not identical.

Let me start with an area of agreement. There is one broad and basic goal of science on which scientists and society probably agree, possibly because this objective is virtually a definition of what science is. This is the objective of increasing our understanding of the nature of the universe in which we live. This we do as scientists by experimentation and by theorizing and by more experimentation and by more theorizing. This is how science has been advanced, through successive stages of progress and study, toward its great goal. This scientific effort, which we sometimes call basic research, is a tremendous and continuing intellectual challenge, and the important breakthroughs in understanding are exciting and satisfying to scientists and laymen alike. Thus, the identification of the role of DNA in heredity or the determination of the age of the moon by analysis of moon rocks not only receives top billing at scientific conferences but also rates front-page coverage in the daily newspapers.

This basic goal of science, to understand the world, leads directly to the other principal expectation which society has had for science: that science would contribute to the modification of the world in ways which respond to the desires and interests of man, or at least respond to those of certain groups of men. And this expectation is very sensible and plausible. Clearly, to understand the nature of things is an important first step toward their manipulation or modification.

The name technology is given to this application of knowledge for the modification of our world in ways we desire. Indeed, to much of society, science has as its principal objective to be the handmaiden of technology, that is, to help out in this modification of the world. Thus, to much of the world it is science which produces radio and television and which conquers space by airplanes and satellites. And if you try to correct this notion by saying: "No it is technology which does that," you can expect an indifferent shrug of agreement. The fact is, for good or for bad, the concepts of science and technology are inextricably linked in most people's minds. Thus, when people want more guns or more butter, they will inevitably turn to science and expect science to help. To society, this linkage of science to the accomplishment of the world's work is an overwhelmingly important fact. Here then is a key objective which society has for science.

Society on other occasions expects additional things of science. Prestige is one; pleasure is perhaps another. After all, the television story of the Apollo Mission to the moon was great fun to millions, me included. And even though a Nobel Prize may rate in national recognition somewhat below a record-breaking track mark at the instant of attainment, the staying power and prestige of the Nobel Prize may be a good deal greater.

To many of the developing nations of the world the view of the objectives of science just outlined is a fully satisfying one. But in the developed nations and in the United States in particular, society is seriously rethinking its objectives for science, and we as scientists cannot escape involvement. Hence, this symposium.

What else does society, our society, want from science? Principally what it wants is *help in the utilization of technology.* This is the big new change in attitude. No longer do we see just more technology as the answer to our problems. Increasingly we see that technology has effects which often go beyond its primary purposes. We are troubled by increased urbanization, clogged transportation systems, pollution, poverty in the midst of plenty. Society increasingly asks for better ways to *manage* technology and turns to science for help.

Society also asks for help in *controlling and restraining technology.* In many areas technological progress has ceased appearing benign and carries instead the image of a remorseless juggernaut that is almost unrestrained and beyond human control. For example, we seem destined to watch helplessly as the supersonic transport carries us down the path to still more noise and shock in an already noisy world. Or to take what may be the most troublesome example of all, the entire world appears to be locked into a technology-based military arms race which grows increasingly costly and which, more often than not, appears to decrease the security of men and nations rather than to increase it. The question here and in other areas is: What can we do about it?

Here then are the new questions that society puts to science: How can we manage the technology we want and need, to maximize the positive benefits and yet to minimize the negative? And how can we effectively restrain those technologies that appear out of control?

Knowing that society has new objectives for science, we can turn to our second question: What are scientists for? And clearly it is not adequate just to say that they are to produce science. Most scientists have always been something more than purely scientists. Some scientists are also educators; others are partly managers; and still others are partly industrial or governmental technicians involved in generating or analyzing new technology. The number of scientists devoted solely to the production of basic research has always been small and will almost surely always remain small. To answer the question properly of what scientists are for, one must consider, along with the science they practice, the many other activities that engage them: education, administration, industry, government, program planning, and so forth. So our question should really be restated: What must be added to this list of activities if scientists are to respond to society's newly recognized objectives?

A first general answer is more *involvement.* A second general answer is more *formal* concern with the applications of science and with multidisciplinary approaches to the solutions of the problems that these applications occasionally cause. However, these statements are quite general and I will be more specific.

First, as to involvement. Scientists must increasingly recognize that science is *used.* Often it is used for good; occasionally, for bad. Furthermore, society's coupling of science with technology inescapably means that judgments on misused technology will inevitably reflect on science. Hence, for the good of their own professional field, along with other more admirable reasons, scientists must study the social consequences of science and technology and occasionally be prepared to stand up and be counted. So also must scientists analyze the even more sticky problem of science and values. What should be our values, and how may science help toward achieving them? As they study these social and value problems, almost surely scientists will be pulled into the programs of social action which these studies suggest.

Perhaps the most salutory lesson from the great ABM debate of last spring was the demonstration that a proper consideration of major technically based decisions absolutely requires a clear and full explanation of the scientific and technical aspects of the issues. Without this, decisions would be based on hardly anything but faith. And so it also goes similarly with problems of pollution, of transportation, of urban congestion, of population control. Willy-nilly, it is safe to predict that scientists in increasing numbers will be impelled toward concern and involvement with the implications of science.

The second specific is that scientists must increasingly expect to participate as *contributors* to multidisciplinary programs of teaching, of research, and ultimately of social action. The problems that confront and worry society do not fall neatly into subject matter packages; they are persistently interdisciplinary and often involve political matters as well. Furthermore, and most important, both the analyses and action programs require continuing interaction among many disciplines. The engineer and the lawyer cannot just talk once; they must stay in steady interactive contact in the study of a problem. Procedures for interdisciplinary analysis and planning must be learned and must be taught, and scientists must be involved. This requires new university programs of teaching and research; it requires new kinds of graduate training; it requires new organizational structures. In other words, if we scientists are to participate seriously and meaningfully in the important problems of society, we must work to change others and must expect ourselves to be changed.

As we contemplate this reorientation of the efforts of scientists, we must ask two further questions: Where and how much? On the second of these, the important thing to stress is the sense of *urgency* that goes with society's new objectives. To solve our problems of peace, of poverty, of pollution, of population, we have so little time. Not all, or even most, scientists will turn to these urgent applied problems. But if we are to have a reasonable chance to solve them, many scientists will need to become involved. As John Platt has recently argued, we must have a truly massive effort.

Where shall we do these studies? Many people argue that the university cannot change far enough and fast enough to respond to these new objectives, and that the needed new multidisciplinary, problem-oriented efforts require the establishment of new institutions—as for example a major new institute for environmental studies—to house and support them. The case for several such new organizations is a persuasive one. But beyond this there are many who feel that the university is poorly suited for developing the required interdisciplinary approaches, the argument being that a problem-oriented program is anathema to the nature of the university.

I want to argue, however, that the university cannot escape involvement and must actually be given an important role. The needed groups of scholars and technicians are already located in the university, and, above all, this is where the students are. Surely it is worth a major effort in restructuring the university to be sure that the faculty talents are fully available for these new objectives and especially to ensure that college students have a full opportunity to learn about the new goals and problems and become skilled in the knowledge and tactics needed in the search for new solutions. If the experience with Cornell's new Program on Science, Technology, and Society is to be in any way typical, students will, in fact, respond to these new objectives and so will the faculty. In no sense is this argument directed against basic research. Basic research is the key to the knowledge that we must have to reach our long range goals. But scientists must become very much more involved in the problems of society. I am absolutely persuaded that both this and basic research should and will go on.

Now let me summarize. Society, I submit, is rapidly modifying the objectives it holds for science. It wants more help in its technology. It urgently needs help in solving the great problems of peace, poverty, pollution, and population. Scientists cannot fail to respond. And the response will involve much more social involvement by scientists and greatly intensified participation of scientists in problem-oriented, interdisciplinary efforts.

Will we succeed? Will we be useful? In thinking about this, I recall a brief scene from the comic strip, *Pogo*. The main characters of the Pogo group are concerned with the large serious problems of a fire and its aftermath, so that Pogo and Alligator and the others are deeply occupied in a big business fashion. In each of the scenes of this skit, down in the lower left-hand corner of each drawing, is a little rabbit who has a fire hose, and his steady response, given as an aside to the activities of the big boys, is: "And I carried the fire hose." In the climactic strip, he even gets to use it. Maybe that's a proper new objective for scientists in responding to these major social problems. Perhaps we can carry the fire hose, and maybe once in a while even get to use it.

Chairman:
One month before this conference, the article "What We Must Do" by John Platt appeared in the pages of *Science*, the journal of the American Association for the Advancement of Science. This article has been very widely received and is the topic of much discussion. In this article, Dr. Platt estimates the half-life for the destruction of mankind that can be expected from existing circumstances like the likelihood of nuclear war in the world situation today. He comes up with a shockingly small figure. He points out that we should determine our priorities on the basis of these findings, and we should work on solving existing crises in the proper order. He calls for a massive mobilization of scientific manpower and designates both basic research and the man in space program as overstudied, perhaps even a waste of time in view of the shortage of time.

Science and the University Crisis
John Platt

To some degree, I shall be repeating here the arguments in my article "What We Must Do."[1]

The situation in the world today needs to be understood on a large scale in historical perspective. It was Abraham Lincoln who said, "If we know where we are and where we are going, we can better understand what we must do." The first thing for us to understand is where we are and where we are going.

We all need to realize in a general way the great changes that have taken place in the last hundred years. Some of the most important parameters of human communication and interaction, for the world system, show enormous exponential growth over the last hundred years. Some of the parameters have grown slower

1. *Science,* 28 November 1969, pp. 1115-1121.

than others, but some have experienced almost stepwise growth. Thus, the change in speed of communications around the world, in going from the velocity of horses, trains, and ships to the speed of light (for radio and television) is an increase by a factor of about 10^7. The increase of travel speed, in the transition from travel by ships, horses, and railroads to travel by jet planes, or to the SST that might soon be made in Europe, is by a factor of 10^2.

In the year 1940, one of the major concerns in the world was what to do for energy when our coal supplies run out—supplies that may be depleted within a few hundred years. But suddenly, since 1945, we have found a new world of energy, from uranium and thorium, and eventually fusion energy—energy enough for millions of years at the American rate of energy consumption. The energy increase just in the uranium and thorium reserves is of the order of 10^3.

In the case of weapons, we have changed from the 20-ton "block-buster" of World War II to the 20,000-ton bomb at Hiroshima, and now to the 20 megatons available in hydrogen weapons. This is a factor of 10^6 in the last thirty years! Likewise, since the first electric computers, the change in rate of data handling has been a factor of 10^6 or thereabouts.

Numbers can also be put on things which are less technological. Our ability to manage diseases, at least bacterial and viral diseases, has increased enormously. Decreases in infant death rates and morbidities from these diseases are difficult to make quantitative for widely varying social and population structures, but I think a number of the order of 10^2 might be appropriate. In the case of the population explosion the increase in the rate of growth during the last century has been by only a factor of 10. But it is interesting to note that the population "doubling-time" in paleolithic times is estimated to have been 30,000 years. The doubling-time is estimated to be 30 years, or something like a thousand times the rate of population growth in prehistoric times.

One can continue in this fashion to more qualitative changes in the last hundred years. In the exploration of the earth, for example, a century ago it took Richard Burton with his enormous pack trains several years to search for the source of the Nile. Today, satellites fly over this area every two hours and can photograph not only rivers but trees and houses and everything down to the size of a golf ball.

In short, the large exploration of the earth is now at an end. Man has now lived at the North and South Poles with hot and cold running water, helicopter service, and nuclear power. In fact, the age of evolution is at an end, 3 billion years of it—evolution, that is, by *natural* selection. It is now *our* pollution, *our* preservation, *our* breeding, or *our* protection which determines the numbers and kinds of species of animals and plants all over the world. Evolution from now on will be by *human* selection, whether intended or not. DDT is found in Antarctic penguins. Only three thousand great whales are left in the Atlantic, and they will survive for another two or three years at the present rates of killing. There are only three thousand tigers left in India. Even far from human settlements, it is man, consciously or unconsciously, who is determining the densities of all forms of biological life on the globe.

What we see altogether, therefore, is a fantastic pattern of change —changes not by a few percent or by a few hundred percent, but by *orders of magnitude.* There have never been changes of this degree in the history of mankind, in scale or speed. Our existing organizations are not designed to deal with changes of this magnitude. And this is the real origin of our current unrest and of the current crises which confront us.

On the other hand, it is interesting to realize that in spite of these enormous changes, there are many cases where a leveling off is being approached; that is, many of these processes which have experienced exponential growth are now beginning to approach their physical, or natural, or biological limits. To give some examples: In communication, we are already operating at the speed of light—and no one expects an increase by another factor of 10^7 in the next hundred years! In speeds of travel, we are traveling so fast that we are now in orbit flying right out of the atmosphere. (When I say "we," I am thinking of us all as taxpayers.) In energy, we have swum into this fantastic sea of almost infinite energy. Another such change to another sea, with still more infinite energy, would not make very much difference in our lives.

In weapons we have swum into an area of overkill, where the individual weapons are too large even for military purposes, at least for maximum area destruction. And today the total overkill capability is enough to destroy all life on earth down to the cockroach level, if all the nuclear weapons were exploded. In short, we already possess the "doomsday machine." In data handling, we are beginning to approach the limits set by signal travel times between parts of the computer, at the speed of light. There may be another growth factor of 10^2 or so from miniaturization, but no one anticipates another extension by 10^6 in the next thirty years. In the case of diseases, essentially all our viral and bacterial diseases are potentially under control, except perhaps for cancer (and it is not yet clear whether cancer is or is not a viral disease). In the case of population, we are within a time that might be as short as thirty years—and in any case not more than two hundred years—from some kind of leveling off. This leveling off can result from reaching the population limit set by foodstuffs; or because we wipe ourselves out; or because we have leveled off growth at some sensible and more humane intermediate population that we have deliberately chosen.

The result of this great growth, and this imminent leveling, is that we are within sight of a new world, which might be a kind of "steady state" world. After men had domesticated horses, their new speeds of military affairs and communications were retained for thousands of years. So today, if we are able to manage these new powers and densities of interaction, we might be able to build a society that would keep us alive for thousands or even millions of years into the future.

The only question is whether we *will* be able to build, and will build, such a society *in time*. In order to grasp the problem, we have to understand what the time scale is. The time scale is set by the rates of exponential growth already discussed, these rates which are just beginning to overwhelm our human institutions. Today the human race is at a stage somewhat like a rocket on a launching pad which is suddenly fired, producing exponentially increased power and stresses that the parts of the rocket have never experienced before. Under the new load the rocket suddenly begins to vibrate and shake, with the danger of blowing apart on the launching pad. Yet, if the rocket can get off the launching pad, it will fly on a new and different course for a long time to come.

It is the next few years, as our social institutions continue to encounter these increasing order-of-magnitude stresses and vibrations, that are the crucial times for the survival of the human race. I would say that our crucial time for decision and survival is the next ten to thirty years. My own opinion is that, because of our still uncontrolled danger of nuclear escalation, our half-life may be in the range of ten to twenty years or less. The reason is that our various exponential crises, as they come together more and more often, will exhaust us, make us despair, and leave us with our mediators worn out, and our administrators facing new complex problems that they have no idea how to cope with. Because of these multiplying crises, I think the time, in fact, is very short and that we may have less than a fifty-fifty chance of living until 1980. I have not talked to any scientist who has considered these problems in some detail—the whole sweep of them—who estimates a very much longer time. U Thant and Robert McNamara and C. P. Snow have all estimated that we have less than twenty years to solve our problems; and U Thant says only the next ten years.

Yet the reason for painting such a black picture is not to say that it will *certainly* come to pass. After having given this picture of the scale and urgency of our problems, I would like to try to paint the opposite side by showing what tools we might have available for dealing with these problems.

My view of history is somewhat like that of Harvey Cox, the Protestant theologian. He has commented that there have been three views in history in the past: first, the "apocalyptic view," that we are *doomed;* second, the Messianic or "chiliastic view," that we are *destined;* and the third, and most realistic, the "prophetic view." Cox identifies this latter view with the Old and New Testament prophets whose predictions were *conditional* predictions. They said, *if* you commit adultery with your neighbor's wife, *then* you will roast in hell. On the other hand, *if* you love God with all your heart, *then* you will create heaven on earth. The point is that we are dealing not with objective predictions, but with an *if-then* situation, where the appeal is for action. What I am saying is that our situation today has this short time-constant, *if* we do nothing adequate to deal with it.

Now, what approaches might be adequate for dealing with such enormous powers and problems? These order-of-magnitude changes are changes resulting from science and technology, which are in large measure the result of increased control over nature. They are problems which are not to be dealt with on familiar grounds by the businessman, or the politician, or the labor leader, or the average citizen. Even though these people are men of good will, they are not handling the full transforming power available at the bench of the scientist or the academic in the laboratory, who could be working to find solutions in large-scale terms. And when I say "scientist," I include the social scientist and political scientist. In fact, I do not refer only to scientists in the laboratory or scientists in academic life, but I mean ingenious minds of all kinds who could use analytical thinking and invention to help us—lawyers, teachers, doctors, engineers, people who are willing to think analytically about our problems and who are willing to work on the problems and look for mechanisms of solution that have not yet been arrived at. I would suggest that for these vast new problems, what we need is something like research-and-development teams for what might be called "social inventions"—inventions that are capable of managing technology, of setting up organizations that are able to deal with technology, of providing feedback and stabilization devices to keep us from blowing ourselves up or tearing ourselves apart because of our inability to control these new relationships.

Recently, Karl Deutsch and I prepared a list of important social inventions made in this century. These are, so to speak, software inventions which change the relations between men or their ability to handle problems. Such an invention would be the work of Keynes in economics (whether you agree with it or not) which represented an analytical attempt to understand and manage the old problem of boom and depression that we encountered up until 1939. In 1929, this would have been *the* crisis problem, and we would not have known how to deal with it. People would and did talk about sending petitions to Washington, they talked about social and political pressures, and so on—but what was actually needed and finally was most effective, was Keynes and a few other men working carefully at a theory of economic structure and economic feedback and mechanisms of stabilization that showed how to manage the problem.

Similarly, in operations analysis one might also identify Blackett and his work. Commercial inventions such as credit cards have greatly changed our relationships through making credit available for recreation and international travel. One might also name "sociotechnical inventions," such as television or oral contraceptives, which have a technical component but which have changed society in a fantastic way. Or one might name legislative inventions, such as the New Deal, which within only about four years completely changed our attitudes toward labor organizations and welfare. One might name international achievements of peace-keeping, such as the Test Ban Treaty, the Non-Proliferation Treaty, and others of this sort. Each of these social inventions was put forward by one man or a few men who were doing analytical thinking, analytical research on the nature of the social problem.

It is surprising how fast these social inventions have taken effect. Our accepted notions of "social lag" are often very mistaken. The time from the publication of Keynes's book in 1936 to the time of the general adoption of Keynesian principles to some degree by both Great Britain and the United States was about seventeen years. In operations analysis, the time from Blackett's study of the submarine problems to the time when this study made an order-of-magnitude change in the ability to cope with submarine warfare was about one or two years. In game theory, the time from von Neumann and Morgenstern's book in 1946 to the adoption of game theory ("zero-sum" game theory, unfortunately) by strategic analysts at Rand and in the Air Force, was about seven years. In the case of television and oral contraceptives, times in the range of ten to fifteen years elapsed before they swept across the country. The New Deal time was about four years. The pay-as-you-go income tax changed the taxing power of the federal government by a factor of five within about two years. (This was not necessarily a desirable thing in general, although at that time when there was a war against the Nazis it seemed like a highly desirable thing for the whole country.) The reason this plan changed the taxing power so greatly was because in the bad old days, on March 15, they could only take what we had at the moment, and this amounted to 3 to 5 percent, perhaps, of the total year's income. Now in the "good new days," they take it away from us before we see it, and the tax rate is more like 20 to 30 percent. I am not advocating pay-as-you-go income tax; rather, my purpose is to point out that a single man, Beardsley Ruml, thinking about how the government could meet its wartime responsibilities, came up with an idea which totally changed by almost an order of magnitude the ability of the government to deal with social problems, in one fell swoop. Similarly, the Test Ban Treaty was first proposed publicly by David Inglis in the *Bulletin of the Atomic Scientists* in 1953. It was adopted in 1963, thanks to Linus Pauling and all of his scientists' signatures, after only ten years.

The net result is that the mean time for these social inventions, from the time of conception or publication, to the time when they exercised a large-scale effect on society, was only about twelve to fourteen years. This is a time comparable to the time of acceptance of technological innovation.

We see that we really live in a very responsive society. I emphasize this because although the available time before we are overwhelmed by some of the coming crises may indeed be very, very short, it may not be too short for groups of people, thinking analytically about how to improve our situation, to have large-scale effects on the outcome and to increase the probability of our survival by some appreciable amount.

Now to say something specifically on the role of the university. Many of the software inventions mentioned above were invented in the vicinity of the university. Game theory came from the Institute for Advanced Study at Princeton, Keynesian economics came from Cambridge, England. Operations analysis was invented by groups of nuclear physicists and linguists and cryptographers. The Test Ban Treaty came in some measure from Chicago, Caltech, and the Pugwash Conferences which were a kind of intellectual concentration. The pay-as-you-go income tax was proposed by Beardsley Ruml, who was the dean of social sciences at the University of Chicago.

Universities and intellectual centers, such as Cambridge and Princeton, have had a role out of all proportion to the simple number of scientific or analytical thinkers on their staffs. The combination of ideas, the resources, the library and laboratory facilities at these centers are conducive to thinking in fundamental and large-scale ways.

Of course, millions of people who are not scientists are already working to solve some of our problems—the millions of people in labor unions, citizens, housewives, neighborhood groups, social workers, businessmen, and others. (It also seems sometimes that other millions of people are working to make our problems worse!) But in addition to these millions, if we are to save ourselves, I think we must also put into this crucial battle for survival a substantial number of basic research scientists. And many of the best ones will need to turn away from what they are doing in long-range research to something which is more urgent and more important, hopefully in time to be effective. Operations analysis in England had to be done by Blackett, who turned aside from nuclear physics, and in this country by such men as Morse and Kimball who turned aside from chemistry. Game theory was done by a quantum theorist, von Neumann, who turned aside to work on economic problems. What is important is not so much the discipline you were trained in, as the fact that you bring the resources of analytical thinking to these problems. And this is something that the universities particularly have to offer.

My own opinion is that within the next year or two we ought to find ways to help something like 10 percent of our basic scientists get into these efforts towards looking for the deep, long-range solutions to our social crisis problems. Ten percent of our basic scientists would mean a number of the order of ten thousand scientists. This would require a total expenditure of the order of $1 billion in research. (This allows each scientist to have a salary of $100,000, of course, but I hope this will pay for a few assistants and library facilities as well.) This sum is comparable to the annual amount spent at NIH, and a sum comparable to that used by the old NDRC in World War II. I think it would be perfectly appropriate to spend on our social crisis problems for this number of scientific brains a sum comparable to the amount of money that we spend on NIH. I am not belittling NIH in any way; some of its research is, in fact, most relevant to the urgent problems of food, population control, and survival. But I am saying that in addition to what we are now spending on NIH we ought to be able to put another $1 billion or so in research on these other human crisis problems. A billion dollars is the difference between putting men on the moon four times a year at $4 billion, and putting men on the moon three times a year at $3 billion.

This kind of endeavor to respond to our social crises could be of considerable value to university education, and even to the faculties who now in many cases feel that they are doing irrelevant work. If a national emergency, like the war against Hitler, were to occur again, there is no doubt that hordes of our most dedicated scientists would suddenly be eager to postpone their basic research for several years in order to do the things that need to be done more urgently. I think the present human emergency of future survival all over the world calls for this kind of postponement just as intensely, in fact more intensely, than any past national emergency. The young see this. In fact, I have never discussed this problem in front of university students without having dozens of them exclaim, "That's what we want to work on! That's what we are here for! Where do we sign up?" And I have not been able to tell them where to sign up.

I also think that this direction of research into urgent problems would be of great importance for getting greater support of science and of universities from the public. In the case of polio, when the research supported by NIH and the Polio Foundation led to breakthroughs in vaccines and cure of the disease, the public was enthusiastic about this success and the support of health research increased enormously. Yet polio is only a second-order disease compared to the problems of food, of hunger, of population control, and of cancer. Today, much of our academic research effort is poured into tinier and tinier studies that may take thirty years to come to fruition. If instead some more human-oriented science and development could show the public in a convincing way that we are working on things that have three-year scales of urgencies—things that may increase the probability of survival, things that are centrally important such as food and population—then I think that the level of public support for scientists both by our state and national bodies also would suddenly increase.

In short, I think there is nothing but gain, both for the human race and for the scientific enterprise, for the universities to begin to apply their knowledge and resources more directly and more immediately to the solution of the problems that have the potential of killing us all in the next few years.

Chairman:
George Wald is professor of biology at Harvard University. He is a successful scientist, as compactly demonstrated by the fact that he received the Nobel Prize for medicine and physiology. He has received other scientific and humanitarian awards, too, some quite recent and rewarding. Professor Wald has written many scientific articles but perhaps none equal in importance, acclaim, or impact as the one published on March 4, 1969, entitled, "Generation in Search of a Future." Reprint requests to the Boston Globe **have exceeded 85,000. It was an unforgettable experience for the privileged few that witnessed that speech** live, and the significance of the event has yet to be realized.

Problems and Responsibilities
George Wald

Professor Long set forth our problems and asked: "What is science? What are scientists for?" John Platt has just suggested an approach to some solutions. As he says, we haven't much time. But we *had* a lot of time! There is a kind of hysterical quality to much of the discussion now. One does not need to be surprised at that, for things are coming at us too fast—too many things too fast. A while back we seem to have gone off the track. We are in a tremendous crisis of confidence— a crisis of belief, a crisis of meaning. Professor Long asked us, "What are scientists for?" We had better ask, What are *men* for? What is *life* for? We have gone off the track, so I am going to try to give you some religion. The only hope for us in a way is that we have gone off the track so badly so recently; it gives us hope that maybe it is not yet too late to get back on it.

About a century ago, not much longer, scientists began to whittle away at man's traditional views of himself and his place in the universe, and for a long time substituted nothing for them. Now we have to reach out for a credible basis for belief in the dignity of man and the sanctity of human life—indeed, in the sanctity of life. I know no other place to look for that basis than in science, so let me just try very briefly and superficially to describe this.

We know by now, to a degree we never dreamed we would reach so soon, that the universe has a unity. It is a historical universe, in which not only living things but stars and galaxies are born, come to maturity, grow old, and die. In the universe, life in each place comes as a late event. Given enough time and given proper conditions, life is inevitable. It is part of the order of nature—a cosmic phenomenon. Living things—the ones we know, but I am sure all others wherever they may occur—have the same fundamental constitution. They are made almost entirely of four of the ninety-two natural elements: hydrogen, carbon, nitrogen, and oxygen.

The universe—if not the whole of it surely large parts of it—was born of hydrogen. Those other elements had to be made, and we have an idea now how and where they are made. They are made in the deep interiors of dying stars—the so-called red giants. It is the red giants that cook in their deep interiors—at temperatures of about 100 million degrees—carbon, nitrogen, and oxygen. These elements are then spewed out to become part of the great masses of gases and dust that fill the universe. Here and there in that gas and dust an eddy forms, a new star is born—a later generation star, not one of the first generation that had to be made wholly of hydrogen. A later generation star, such as our sun, has the carbon, nitrogen, and oxygen in its structure. It is in the planets of such stars that life can occur—given enough time, I think inevitably occurs.

Life is a great thing in the universe. It is the most complex state of organization—so far as we know or can even imagine—that matter can assume in our universe. How much life is there in the universe? Well, as far as we can tell, lots of it. There are about 3.5 billion people on earth and a lot of us are beginning to feel crowded. But there are 100 billion stars like our sun in just our home galaxy—the Milky Way—something that now seems as cozy and homey as the front yard, yet it contains 100 billion stars! What a curious bit of arrogance to think that unidentified flying objects represent visitors from high technological civilizations in outer space that have come to look us over. Why would they come here? They have 100 billion solar systems to choose from. Why come here? Utter nonsense.

So there we are. The present estimate is that of the 100 billion stars in our own Milky Way, roughly 1 percent might have climates capable of bearing life, and many of them should do so. A billion places in our own Milky Way! There are already perhaps a billion such galaxies within reach of the most powerful telescope —the 200-inch facility on Mount Palomar. There is a lot of place for life in this universe and undoubtedly a lot of it there.

Life is a great thing. The bigger a star is, the faster it consumes its hydrogen, the faster it comes to the end of its period on the main sequence and begins to die. Our own sun is middle-aged. It has had about 6 billion years on the main sequence, and we estimate that it has another 5 or 6 billion years to run. A star with twice the mass of the sun completes its period of maturity, its period on the main sequence, in only about 2 billion years. Life has existed on this planet for longer than that—about 3 billion years. Many a star has been born and died since life arose upon earth. A star with twenty times the mass of the sun—and there are such stars—completes its time on the main sequence in about 3 million years. That is only as long as manlike creatures have roamed the earth—and that brings us to man.

Life is a cosmic thing. It is a great event in our universe and part of the order of nature. Man and his like are a great thing. You may have heard it said, a hen is only an egg's way of making another egg. Well, in just the same sense, a man is the atoms' way of knowing about atoms. Those atoms that constitute man had to wait a long time, but then it happened. Without such a creature as man, the universe could *be* but not be *known*; and that's a poor thing. So here we are, a kind of culmination. Man, the dominant species on the earth coming at the end of 3 billion years of evolution, made of stuff picked up from every corner of our galaxy over endless reaches of time—the stuff of dying stars. That is what makes a man, and that is what makes life.

Even biologists are in the habit of thinking of the environment as something *given,* and that life has to find its place in that fixed environment. It is the environment, we are told, that plays the tune, and it is for life to either dance or die. But it really is not that way at all. Some of the most important features of our physical environment are the work of life. One such feature is the oxygen in our atmosphere. There was no oxygen gas in our atmosphere until it was put there by life—until it was put there by photosynthesis performed by plants. Right now, our whole earth is in a delicate balance maintained entirely by life. Every bit of oxygen in our atmosphere comes out of photosynthesis and cycles back into cellular respiration—a cycle that completely renews it every two thousand years. Two thousand years is only a day in geological time. With carbon dioxide it is stranger still. Every bit of carbon dioxide in our atmosphere and dissolved in all the waters of the earth comes out of cellular respiration and goes back into photosynthesis every *three hundred years.* Every bit of water on the earth goes in and out of living organisms and is completely renewed every 2 million years. And what is 2 million years? When you think about our dependence on the seas you think of fish, and some who do not care much for fish would rather disparage that dependence. But please remember, it is only life that keeps oxygen in our atmosphere by photosynthesis, and seven to nine tenths of all photosynthesis is done in the surface layers of the ocean—not on land where we are familiar with it, but in the surface layers of the ocean. Polluting the ocean enough will put us in far worse trouble than not having fish on Fridays.

So here we stand as products of a long and tremendous history and suddenly find ourselves at a kind of parting of the ways, a kind of crossroads in which one has to make the choice—and very quickly—whether it is to be life or death. Where does the scientist stand in that decision? First of all, that man—who is the atoms' way of knowing about atoms, who is the stars' way of knowing about stars—that man is a scientist. What is the world to look to him for? Certainly not just for the roots of new technology, but as the man who *knows* and who has now to serve as priest for mankind. There was a time when *priests* were scientists. Now that priests have stopped knowing science, it becomes necessary for scientists to become priests. It is important now that the scientist, in being the man who knows and is deeply concerned with that knowledge, and deeply concerned with the future of the human enterprise, is something of a priest in his outlook. It is perhaps most important that he be disinterested, that he ask nothing for himself. These things too are what scientists are for.

We are at a parting of the ways not only for humanity but for much of life on the earth, all of it now in our custody. Things live or die on this earth now, depending upon what man permits. That is the choice. Though I look to science for most of my religion—mine is a wholly secular religion—yet I go to the Bible for my tradition. And so these days, all the time, there run through my head those beautiful words in Deuteronomy, chapter 30, verse 19: "I have set before you life and death, blessing and curse; therefore choose life, that you and your descendants may live."

Chairman:
Victor Weisskopf is institute professor and chairman of the Department of Physics at the Massachusetts Institute of Technology. He is a quantum physicist who has made major contributions to our understanding of atomic and nuclear structure and more generally to the foundations of physics. He is an accomplished author, scholar, teacher, and lecturer.

His scientific career began in the late twenties in the golden age that saw the birth of quantum mechanics. Included among his early associates and colleagues are Pauli, Wigner, Bethe, and Niels Bohr. Professor Weisskopf has proved to be an articulate spokesman in support of the need for basic research, especially basic research in physics, and perhaps more especially in nuclear and high energy physics. He was at Los Alamos at that time and has served as director of CERN. Both MIT and the greater Boston scientific community are pleased he returned from CERN, for MIT is now grappling with some fundamental problems that involve the socially concerned student, the desire and the ability of the university to do research, the United States government, and mutual priorities. Dr. Weisskopf has been a central figure in these considerations at MIT, and his recent experiences and concerns and thoughts will undoubtedly surface in his remarks to follow.

Why Basic Research?
Victor F. Weisskopf

Basic science is today in a very difficult position. It is attacked by what I would call the Right and by the Left. The Right, conservative circles, would say that basic science must be useful. It is a luxury if it is not useful, a luxury which we cannot afford. Science must show its "rentability," its payoff in terms of adding to our health and profits, or providing better weapons. On the other side, the Left (for lack of a better name) which includes a good part of the younger generation, says that basic science must be distrusted.

It was, is, and will continue to be the source of innovation—technical innovation which will be used by industry—that will continue further to deteriorate our environment. It has created and will create social problems. It will bring us nearer to Orwell's *1984* and, last but not least, it will create worse and more deadly weapons. At best, they say, basic science is a waste of resources that could be much better used for socially useful purposes.

Basic science finds itself between these two wedges, and basic science may disappear. Support for it is already reduced and will be reduced more. Demoralization will set in among those people who profess it. This demoralization comes not only from the decrease in financial support but also from the attitude of the public which expresses distrust and lack of interest. The source of basic science may dry up; for only a small number of people, as Franklin Long pointed out, are actually engaged in it as a profession. The supply can easily run out, especially if the students no longer come to it, if they no longer are enthusiastic about it and choose to profess it as their study.

Is this situation good or is it bad? Let me say that the state of basic science poses a particularly great danger, particularly now. We are living in critical times, as all of our speakers point out. What we have here is a problem of *time* and a problem of *space*. The problem of time—as John Platt showed us—is that the development of all behavior patterns affecting our way of life throughout the world has gone faster and faster. We are now at a critical stage in the development of mankind where changes are considerable and qualitative even within one generation! This means that we can no longer learn what to do from the lives of our parents. This is a new phenomenon. The space problem is that the changes have become so great and so all encompassing that they include the whole earth. The earth, or space in general, is no longer infinite. We can no longer neglect the effects of what we do on the earth, for we are changing the planet. These two problems are new, as John Platt explained. They have occurred only in the last few decades, and this is the source of our crisis: *we are facing completely new problems.* The effects of all that happens here must be thoroughly studied, and for this we need pure science. We will need, in fact, more basic science and more pure scientists than now available. We can get these people only from a continuous and vigorous pure research activity, because you cannot train such people out of books. A quotation from Polanyi makes this point very clearly:[1]

"Encircled today by the crude utilitarianism of the Philistine and the ideological utilitarianism of the modern revolutionary movement, the love of pure science may falter and die and if this sentiment were lost, the cultivation of science would lose the only driving force which would guide it towards the achievement of true scientific value. The opinion is widespread that the cultivation of science would always be continued for the sake of its practical advantages. This is not so. The scientific method was devised precisely for the purpose of elucidating the nature of things under more carefully controlled conditions, and by more rigorous criteria than our present situations created by practical problems. These conditions and criteria can be discovered only by taking a pure scientific interest in the matter which again can exist only in minds educated in the application of scientific value. Such sensibility cannot be switched on and off at will for purposes alien to its inherent passion. No important discovery can be made in science by anyone who doesn't believe that science is important, indeed, supremely important, in itself."

Well, why do we need pure science? Why do we need the attitude of those people who profess it? In response, I would like to separate the reasons into two groups. The first contains the direct reasons. We face problems that are difficult, complex, and new. These are not engineering problems in the sense that we have no

1. Michael Polanyi, *Personal Knowledge* (Chicago: University of Chicago Press, 1958), p. 182.

set rules. We will have to deal with completely new situations, just as we find it usually in basic science where we face completely new phenomena. The problems of environment, pollution, food, and medicine cannot be studied with preconceived ideas. One must be free of preconceived ideas. In other words, one should have the attitude that is fostered and engendered in basic research. This is why John Platt, and rightly so, asks basic scientists to do this work, to join him and others in attacking our problems. They should do it, and I hope they will do it in large number. But we must be careful that in this operation we do not kill the basic science establishment, and it is a very tenuous one because there are not many people who are in it. The trouble is that the crisis we face will last a long time—my estimate is at least twenty-five years, that is, a generation, probably more. Therefore, we need to renew the young generation and we need young people who are trained in objective scientific methods.

Think of Word War II. For us, World War II lasted only three or four years. At that time this country set forth practically all of its basic research capability to solve very definite problems. These were much, much simpler problems than the problem we now face, because those were purely technical problems—radar and the atomic bomb, unfortunately. Now consider Germany. In Germany basic science was cut off not for three or four years but for thirteen years. No science was under attack during the Nazi regime and, of course, during the war. The effects are plainly visible. After twenty-five years postwar, German science is still not where it should be, and Germany is a relatively small country helped by the United States and others to build up its science. Nobody will help us.

We are facing complicated and many-faced problems. It is not like the Manhattan Project. We need the depth; we need the purely scientific interest, as this quotation says. As an example, consider the problem of radiation damage. During the war and shortly thereafter, many people were worried about radiation damage and the effect of radioactive exposure on the body of man and animal. It required the whole development of molecular biology, undertaken in the decades before and after the war and done completely independently of any practical application, to realize the depth of the problem: that radiation damages not only organs but also the genetic heritage of mankind.

Now I come to the second importance of basic science which may be, in many respects, the more significant. Since I can only sketch my arguments briefly here, I am glad that my friend George Wald has helped in expressing what I want to say. For the activity of basic science fosters and engenders a certain attitude toward the environment, toward man and nature, of which George Wald has spoken so eloquently. Basic science can and should create a morality. I can only remind you of the evolution, just described, from the hydrogen atom to man that has happened in 10^{10} years. This evolution is a unique experiment of nature, and it is now in our hands, whether we want it or not. This is the basis— the deep basis—for the responsibility of man. Basic science can and should create more intimate relations between man and things in nature. It should create an awareness of the laws of nature, of the connection between phenomena, of ecology in the widest sense—how everything depends on everything—of how the universe and the atom are one and coexist. Now people do not realize this, for if they did, as George Wald has said, they would vote more money for conservation, for the environment, for quality, and for social purposes. The public does not consider basic research to be important. It has not been brought near to them. We cannot proceed if we have the most wonderful schemes to solve problems on our drawing boards and the general public does not support us. Basic science can and should create a love of nature, an appreciation of the world in which we live, and show that this world is worth saving.

I would like to repeat a wonderful statement of my colleague Robert Wilson, who before a committee of the Senate had to defend the need for the new accelerator planned near Chicago for basic research. Senator Pastore asked him if this machine would bring anything that would help to defend our country against our enemies. Wilson replied, "absolutely not." Pastore then asked, "Well, then what is the use of it?" And Wilson said, in such a wonderful way, "In that sense this new knowledge has nothing to do directly with defending our country except to help make it worth defending."

In the past we have used nature for our advantage—to a short-sighted advantage. We have taken nature for granted. I would say we have regarded nature as a prostitute. But now we should get married to nature. We should take care of her. And it is not to be a marriage of convenience. It is to be a marriage of love and mutual understanding. Perhaps this is the way in which science can create the spirit and the attitudes necessary for achieving a better quality of life for which we all strive so much. How can science do this? How can we scientists do this? I believe it cannot be done by abolishing or severely curtailing basic science, or by switching all of our activities over to practical applications no matter how useful and socially valuable they may be! Also, we cannot do it by continuing the science establishment as before—which, if I may say so, was in many ways only a factory for producing new results as fast as possible. This is not the way. To proceed we have to become aware of several important things. The first step is to anti-specialize science, to create a greater awareness *among scientists* of the coherence and connections in nature. We have to bring fields together, to show connections between different fields of science. There is no such thing as chemistry, as biology, as astronomy, and so forth. There is only nature and man, and that must be seen, taught, and propagated. As much as possible, we must have interdisciplinary studies. In addition, much more emphasis must be given to the popularization of science on every level. Nonscientists and scientists alike do not understand enough of the nature of science. This was clearly shown in the recent ABM and MIRV debates.

The popularization of science is not just to make basic science more popular and appreciated, although I think this is important; there is much more to it. If scientists would think that to be aware of the deep and simple facts of nature is as important as finding new discoveries and helping to improve social conditions, then nature would be helped. The point is, if you cannot explain a scientific result in simple terms to a layman and show its importance, then you do not understand it yourself. With a more concerted and systematic effort toward popularization, I think science will clear itself and scientists will be clearer on what they really do. Another point about the popularization of basic science is that we must not now go to the public with information only in socially important issues, for example, only to explain how science is important for cleaning up the atmosphere. That would get us nowhere. We must go deeper, perhaps in the way that George Wald has just tried to do. We must show from science and with science how much everything depends on everything, how things have come about, and let the public draw its conclusions. Real change in the world can come about only if people want it themselves, not if they are told what to do.

What we have to do as scientists is to continue being interested in nature, to try to understand it in a deeper and more general and encompassing way. We must help where we can help, and there are so many places where we are needed. John Platt is right in asking a large number of us to help. But we should also be with the young, with the students, and teach and live science as a human activity—as natural philosophy. It will be catching.

Chairman:
It is a difficult task to categorize Lewis Mumford or even assign him a title. While everyone associates him with Harvard University, he modestly but quickly points out that he has no claim to a Harvard designation and would be quite content to have no identifying mark in his by-line. When deservedly referred to as a philosopher or historian, he says that perhaps generalist is better. All of these things apply, but Victor Weisskopf made the appropriate designation: "He is a wise man." Mr. Mumford has written many books and articles of world impact and has just completed a book on the problems of society, particularly as they relate to science. When queried on the impending moon landing, he said: "The prime task of our age is not to conquer space but to overcome the institutionalized rationalities that have sacrificed the values of life to the expansion of power in all its demoralizing and dehumanizing forms."

The Integration of Science and Life
Lewis Mumford

Let me confess to begin with that I am under a slight embarrassment appearing here today; for though this is not the first time I have taken part in a meeting of this association, I am not a member, and the only organization I might qualify for is one that has not yet come into existence: the Association for the Integration of the Sciences and Humanities with Life. The existence of such a society would indicate that the original mechanical world picture outlined in the seventeenth century had been replaced by a more organic model; and unfortunately this transformation has still to take place.

Science has followed in the main a course laid down by the founders of the Royal Society in London; and though these bold minds had been deeply influenced by the thought of Francis Bacon, they overlooked a remarkable passage in *The Advancement* of *Learning* which might well serve as the key to this paper. "Mere Power and Knowledge," Bacon observed there, "exalt human nature but do not bless it. We must gather from the whole store of things such as make for the uses of life." To disclaim any concern for "the uses of life" has I fear been treated by many scientists as a legitimate boast, though a growing number of people, not least among scientists, have begun to realize that it is a serious defect.

If, as a generalist, I have anything to contribute here, it will be to point out some of the things that were left out in the seventeenth-century concept of scientific objectivity and that are still for the most part lacking today. Even scientists who now favor interdisciplinary cooperation and research pursue this aim by promoting intercourse within the sciences rather than by reaching out to include human experiences and activities that resist the application of their deliberately depersonalized methodology. This methodology was entirely adequate for giving an orderly account of the behavior of physical bodies and recurrent events; for planets, cannonballs, and machines, so far as we know, have no subjective life; so nothing important was left out of the seventeenth-century view so long as science confined its attention to the physical world.

Even those who, like Lord Snow, earnestly wish to bring together the sciences and humanities have a one-sided view of such unification; for while Lord Snow rightly believes that the humanities have much to learn from science if they are to remain in touch with some of the most stirring minds of our day, he does not suggest that science, equally, must comprehend and find a way to absorb the immense store of human experience embodied in the humanities. If science is ever to be fully integrated with life, it must allow a place in its own scheme for many nonrepeatable events, historic and biographic, and for sundry expressions of human creativity that cannot be externally observed, quantitatively measured, or mathematically formulated. The "founding fathers" of science, Galileo, Descartes, and Newton, were too ready to abandon this vast subjective realm to purely private cultivation, as if man's inner life did not need the order and discipline and rational cooperation that science was establishing in its own special domain. By sanctioning this divorce, science felt free to go its own way, without respecting any human interests, values, or purposes except those that directly subserved science itself.

When Descartes, in his *Discourse on Method,* proposed to give the Church final authority on all matters that concerned the human soul, many later interpreters believe that he was only cannily suggesting that man had no soul. But what he was really saying was that the soul had no place in the scientific *Weltanschauung;* that the realms of art and religion, of morals and politics, of history and biography were outside the scope of its method. After treating one half of man's life, the subjective half, as irrelevant, indeed virtually nonexistent, science gave its description of the other half that claimed its entire attention the curious designation "objective."

I have traced some of the deplorable consequences of this division in a new book I have just finished, *The Pentagon of Power;* but that is too complex a subject to be even touched on in a few passing sentences. Perhaps the most important consequence was to give excessive authority to the processes of mechanization and quantification and measurement which entered into the construction and exploitation of machines. For only in the world of the machine could science be entirely at home, unembarrassed by conditions and events foreign to its purposes. To be perfect, even as a machine is perfect, is to be without any subjective life whatever, except that which directly furthers science and technics. On those terms, no integration of the sciences, the humanities, and life would be possible.

Science's original concentration on the purely external aspects of existence was doubtless responsible for the swift expansion of technology from the eighteenth century on. So striking were the rewards of this method that the positivist avant-garde, by the middle of the nineteenth century, took for granted that mechanization would take command of every activity: that industrialism would soon displace militarism, that scientists and engineers would wield the absolute authority once enjoyed by kings, and that the advances of science would provide immediately on earth what religion had promised only ultimately in heaven. This view was expressed with fervor and unction by the influential minds of the period, Auguste Comte and Herbert Spencer; and though it was flatly contradicted by the rise of military violence and collective human malevolence during the next century, it has become part of the folklore of Western culture. Even now the conviction that science and technology are the only hopeful sources of human salvation has maintained itself in the teeth of massive contrary evidence.

The effort of many scientists still to disclaim any concern with the consequences of their activities has become absurd now that these consequences have become frighteningly visible. And the last attitude permissible in a situation as grave as the present one is that of fatalistic resignation or autistic withdrawal from responsibility. Scientists too often treat the pursuit of science as a categorical imperative, an absolute, whose validity must not in any circumstances be questioned. Too easily they echo the favorite solecism of the man in the street: You can't stop progress, can you? This I submit is a singularly naïve attitude. Scientific progress has been deliberately accelerated in our time by enormous corporate and governmental subventions; and what can be accelerated can also if necessary be brought to a halt. It is time we realized that there is nothing foreordained, nothing automatic, in either scientific or technological progress.

Such genuine progress as has been achieved in any area has been the result of human interests, human decisions, human aspirations. But that progress has likewise been limited, all to often, by psychotic obsessions, by irrational compulsions, by demonic outbursts of destructiveness; and it is precisely because these latter manifestations have now become endemic that the need for an integration of the sciences and humanities with life has become so pressing. If scientific progress still seems to many people irresistible and inevitable, no matter what the actual social conditions that result from it, this is because they have a subjective addiction to automatism, regardless of the results. In this, their behavior is similar to that of our great corporate organizations, such as the telephone companies which, at the very moment their automated system is breaking down through an overload of calls, are still frantically trying in their advertisements to induce more people to use more telephones more often.

All our current plans for science and megatechnics must now face a reaction that almost no one until recently conceived as a possibility: that the very quantitative successes in these areas would offer as great a problem for our own age as ignorance and poverty and physical helplessness in the face of natural forces had done in the past. In the last decade the results of incessant quantification have become too gross and life-threatening to be ignored. Not only is the planetary habitat befouled by mountains of waste and rubbish, by sewage, chemical poisons, nuclear pollutions, but the same choking up of organic activities, the same insidious depletions, are taking place in the mind. Even in science, the uncontrolled quantification of information and knowledge has patently increased the output of rubbish (unusable knowledge) and nonsense (trivial information) and the method of coping with this pollution by means of computers, retrieval systems, microfilms, tape recorders only erupts in larger waste piles, since no effective form of control can be established unless we reduce quantification at its source. This we can do only by the recultivation of human autonomy and measure, by acts of selective evaluation, by deliberate restriction of intake and output to what is humanly assimilable and desirable. Not expansion but Integration is the new order of the day.

Scientists have all too slowly been awakening to this challenge. Fortunately the younger generation has suddenly become aware of the social inertia and the moral contradictions of our society; and they have begun to question not merely the means but the goals that past generations have taken all too glibly for granted.

We in this section of this symposium are somewhat handicapped today by the absence of any vocally protesting students. But if they are absent in flesh, they should nevertheless be present in our minds, for what they are calling attention to by both their rational, well-directed criticisms, and sometimes by their violent and irrational actions, are the weaknesses of a closed system of thought that has no place for any kind of human experience except that which furthers the system.

Perhaps the most serious evidence of disaffection among the young is the fact that so many of them, and among them the most promising, are mentally withdrawing from the great organizations which up to now have claimed their allegiance and their working life. Some of my friends in physics tell me that there are physical laboratories all over the country newly equipped with marvelous instruments for research, mainly prompted by panic over Russia's success with the Sputnik; but these laboratories are now virtually deserted. There are few takers.

Only a handful of students still deign to pursue hard science. How long is objective science going to last if this is the subjective reaction its very success produces? About as long as the Roman Empire lasted, when the Christian dropouts from this society decided they were not having any of it.

Once the power and wealth, the order and drill of the Roman imperium ceased to be morally acceptable or humanly interesting, it was doomed. Institutions are not the result of mechanical and chemical activities operating without human intervention: they are ultimately products of the mind, and they continue to be effective only so long as they awaken positive human responses. If Roman society could collapse and disintegrate once people lost faith in it and turned their minds to a new conception of godhood and followed a new gospel, the same kind of reaction could also be happening in the world we are living in today for much the same reason.

Is this not, indeed, what is actually taking place before our eyes —all the more menacingly because of the subjective disintegration that has accompanied our amazing scientific achievements? If you have followed the art of our time, if you have been to the latest movies, if you have seen the latest exhibitions of the Living Theater, you know that the demoralization and disintegration of our society has already gone far. And unless there is a decisive response to this situation within science itself, its studious indifference to these realities may hasten this disintegration.

Do not misunderstand the point I am making. I regard the scientific method itself as one of the greatest triumphs of the human mind, more important for its moralizing effect than all the technical and practical transformations it has brought about. This method made it possible for qualified observers to reach agreement about processes and events that were open to external observation in such a way that if there were any error it could be detected, and if there were unconscious subjective distortions they would be corrected. As a result there are now large areas of human experience so well defined and so well described by science that no one in his senses would question them. Unfortunately the scientific method cannot be applied so swiftly and directly to the subjective world. Here the need for order, the demand for predictability in conduct, has been satisfied, all too imperfectly, by custom, mores, law, institutional regularities, and has not yet achieved the universality of science.

For this weakness, the founding fathers of modern science are themselves partly to blame. In the conferences that formed the Royal Society in 1664, its members were largely guided by a memorandum of Robert Hooke's. He advised the society not to meddle with "divinity, metaphysics, politics, grammar, rhetoric, or logic." Unfortunately the Society sought to heed this advice all too strictly by exiling, or rather excommunicating, these subjective disciplines without realizing that they could not even have discussed the matter without employing grammar, rhetoric, and logic to come to a decision. These disciplines are creations of man's mind: testimonials to the richness of his subjective life. Nothing that is observable in the physical world would be intelligible without language, grammar, and logic; nor would science have ever overcome the limitations of purely empirical investigation if it had not invented a special language of its own, the language of mathematics, itself an extravagant testimonial to man's autonomous subjective life!

Since the integration of science with life demands a revision of this original illogical bias against subjectivity, I am happy to see that in the many wonderful symposia that have been arranged for this meeting there is one on linguistics. That is obviously a fundamental humanist interest that had been peremptorily excluded from discussion in the seventeenth century. In dismissing this part of man's life, as the founders of the Royal Society did, they were laying up trouble for the world today. Unless we proceed to bridge that gap and to restore intercourse between facts and values, between processes and purposes, that trouble will grow worse.

Even to outline faintly the problem of integrating science with life would require a three-hour lecture, not just another five minutes. So I shall confine myself to a single example of what is implied by integration. It just happens that yesterday I received a singular honor from a society I helped to found: the Society for the History of Technology. I received the Leonardo da Vinci Medal. What made this award particularly precious to me was the fact that Leonardo has always been one of my heroes. His life and thought thread through my work. The thing that makes Leonardo exemplary is something that I have come to appreciate fully only at the end of my own life, while meditating on the meaning of his work; and this is that he was the very picture of an intellectually and emotionally integrated man: not just a scientist, an artist, an engineer, an inventor, or a philosopher, but a man at home in every part of the world. Morally, Leonardo was so sensitive and so alert that he suppressed his invention of the submarine, because he said the human race was too devilish to be entrusted with such an invention.

All through his life Leonardo kept giving attention to every aspect of the world around him: he maintained his interest in art, although he painted very few pictures; he widened his efforts in science, taking in anatomy, geology, and meteorology, but never reached the point of publication; and similarly he was an indefatigable inventor. All the dominant interests of science and technics today were already represented, at least in sample form, in Leonardo's mind. He even anticipated Freud by paying attention to his own dreams, some of which, as I pointed out in *The Myth of the Machine,* have become ominously prophetic in our day.

There are many lessons to be drawn from Leonardo's example; but the most important one, perhaps, is that he demonstrated that the integration of science and the humanities with life is actually possible. This demonstration is all the more pertinent to us today because Leonardo likewise showed that such an integration demands a sacrifice—a willingness to forgo quick rewards in any one department in order to keep the whole range of human activities in better balance. Obviously no one man, not even a Leonardo, can altogether succeed in such an effort: the pressures of his own period, one when despotism was in the ascendence again, certainly caused him to spend too much time in inventing new instruments of war. But Leonardo's abstemiousness, his moral inhibitions, his unwillingness to publicize his work prematurely—he never even made use of the printing press—are the qualities that will prove indispensable in our own day if we seriously attempt the integration of science and life.

This is not, I hope you will realize, a criticism of scientific objectivity as such, but an argument on behalf of a broader approach that will make science capable of accepting every aspect of reality, including what is subjective, internal, nonrepeatable, or visible only through inference from the expressions of art and religion and literature. Once such a unity is established, we shall have taken the first steps toward effecting it in the mind, and making it available for the service of life. That is why the whole life-course of Leonardo serves better as an example of integration than any particular achievement. What seem to be Leonardo's failures—his unfinished paintings, his unedited and unpublished papers, his half-completed inventions—are in fact testimonials to his ability to respond simultaneously to all the challenges of his culture. This contrasts with the unqualified "success" of our power system, whose traumatic triumphs now put all life in danger.

Now if you say that if the world followed Leonardo da Vinci's example a slowdown of activity in every specialized department would follow, you would be correct. There are recent examples in science to emphasize this probable result. Max Born, in his autobiography, confessed that his active personal interest in other areas than physics prevented him from concentrating on that new knowledge which enabled his more famous students and assistants to invent the atom bomb. Precisely. But how much happier the world would be if there were more Max Borns and fewer atom bombs! So far from regarding a probable slowdown as a reason for not following Leonardo's or Born's example, I should say that this is a valid reason and incentive. For only on these terms shall we be able to integrate science with life.

Discussion Session

Question (unidentified speaker): Dr. Long, you raised a serious question about controlling technology. To me, the only problem seems to be that existing political institutions are bent upon controlling man, and they rose in an era in which there was very little technology. I would like to ask, if the problem now is to control technology, don't we need a total shift in the orientation of civilization, whether here or on a global basis, that perhaps would make obsolete all political institutions?

Professor Long: You, of course, realize that my competence does not extend into really knowing the best political structures. I strongly suspect that we shall direct our efforts and act as if we can modify current political structures sufficiently rapidly to accomplish what we want. I don't pretend that this is easy, but the history of mankind, as superficially I am aware of it, indicates that gradual change of social institutions is the most normal state of affairs. I would be pessimistic myself in solutions that involved truly drastic changes in political activity as prerequisite for their success. I am uneasy about putting that much dependence on our political structures for the likelihood of accomplishing our goals.

Question (Professor Carl Sagen, Cornell University): I would like to direct a question to John Platt for whom I have the highest admiration. Dr. Platt in his talk referred to the need for social invention and mentioned a list of social innovations that have occurred over the last few decades. It would seem to me that there is a large-scale inertia both in the public's mind and in the Congress for the very idea of developing such social innovations. Let me point out the great public hysteria on such a comparatively mild innovation as facial hair or the length of hair on the head, to say nothing of recent changes of conduct. I might also mention the general public's hysteria to the innovation represented by the hippies, who truly are a kind of social innovation, although perhaps not a successful one. So, my question is, with the abstemious in-

tellectual habits of Congress, how is it possible to have a billion-
dollar outlay for the development of social innovation, some of
which would surely be, if you will excuse the expression, radical?

Professor Platt: I think that the payoff from the required innovations is a payoff for all of us, and this will affect the only real source
of power in the world that controls the gap between what *is* and
what *might be.* What might be is so fantastically wealthy, with all
our blocks and malfunctions and conflicts removed, that there is
an enormous payoff for the right wing, for the left wing, for the
suburbs, for the ghetto, in solving these problems. An enormous
number of our problems are structural problems, not political problems. Pollution is not a Republican problem or a Democratic·
problem. Improved city management is not a matter of the Left or
the Right. Most of these social inventions that I mentioned—to the
extent that they were regarded as useful inventions—were not pri-
marily political in their adoption. The New Deal is possibly the
worst exception to my comment, but yet it was adopted overwhelmingly at the time. My feeling is that there is an enormous amount
to be done—in the way of solving our structural problems, in cur-
ing pollution, in improving administrative management, in reduc-
ing sources of conflict—which does not require a head-on col-
lision with political forces. This is the business we should get at
as soon as possible.

Professor Wald: You have all heard the phrase, "Praise the Lord
and pass the ammunition." I praise the Lord and am about to pass
the ammunition. Recent discussion and innovation of wearing hair
long and blue jeans with patches—all the rest of it—has not been
taken on and, indeed, has been rather violently rejected by my
generation. The situation is an interesting one. You see, hippies
have a number of troubles, perhaps most of them their own business, but there is one that cuts at the roots of American life—they
don't consume enough! (Applause.)

Before this discussion goes any further, I would like to intrude with an element that hasn't come up yet. Somehow, through most of what has been said so far that involves our practical problems, the assumption is there that if you only know what's good for us, that's what's going to happen. You all know that isn't true. Not in the slightest. The best scientific minds in the country said, "No ABM." But we got an ABM. Was it a scientific decision? It wasn't even a military decision! That ABM probably won't work. So what was it? Twelve billion dollars in defense contracts!

We are facing an escalation in nuclear armaments that is supposed to multiply the supply of nuclear warheads. The present supply, we are told, is capable of killing 140 million Russians, perhaps within an hour. Multiply that by the mid seventies, with the MIRVs, by a factor of five. Problem: There aren't that many Russians. So why do it? First estimate—$17 billion in defense contracts.

We've come to a pass in American life and it spreads from America to the whole world. We mustn't in this discussion neglect any longer the thought that somehow or other you do research and get the social answers. That that view will get you anywhere just is completely absurd at this point. What we are facing is the biggest business in America, the biggest in the world, the biggest that has ever appeared in human history: the present armaments business. You can really almost disregard all questions but the sizes of those contracts. And the crazy thing is that we have become accustomed to them.

I would like you to try to understand for a minute what we are up against. You know, the kids keep yelling at us about imperialism. Ladies and gentlemen, imperialism is old-fashioned, it's old hat. It isn't the real game at all any more, and I will show in an instant how true that is. The gross national product for South Viet Nam is $2.4 billion. If you went there and wrung the country dry as good old-fashioned imperialists, you'd get a cut of $2.4 billion. For heaven's sake, who cares about $2.4 billion nowadays? That war over there is costing us $30 billion a year. The new procurement budget is $21 billion for next year. Do you know what $21 billion is like? The whole of Southeast Asia where those dominoes are, so tick it off: North Viet Nam, South Viet Nam, Laos, Cambodia . . . I always forget one, some Freudian reason no doubt. Thailand. Thank you. The whole of Southeast Asia, those five nations have a total gross national product—all the money that changes hands in the course of a year—of $10.2 billion. And the ABM contracts alone are worth $12 billion! That's the state we are in. Let's just bring that into the picture. (Applause.)

Professor Platt: Does that mean, Dr. Wald, that there is nothing we can do?

Professor Wald: Gee, John, there are a lot of things we can do, but we had better begin to do them. The thought that we will now start doing research, that's fine. Let's do that research, but we are already drowning in information. We know those ABMs are of no damn use. The research had been done—but we get the ABM. We know enough to start doing things, lot's of things. We know enough for many of the most important decisions already. Where knowledge can be improved, by all means let's have people working to get us to know more. But we can't defer action in order to get that information.

Professor Platt: I approve of action but your action apparently is going to be a political action which is a direct confrontation. It doesn't have any of those exponentials in it.

Professor Weisskopf: The worst exponential of all is in weapons development.

Comment (unidentified speaker): I really regret that this program was not broadcast on TV as it should have been. A subject like science and the future of man—that's a pity.

Unfortunately, there are two things I think we are omitting. Except for Brother Wald, there wasn't much talk about the *future* of man. One thing omitted by all the speakers was describing the kind of world we live in today. No one referred to the fact that we live in a world of commerce. We live in a capital society. In order to understand the scientific and all other interrelations and complexities of the nature of the future of man, you must first at least come to grips with what is the kind of world we live in today. And we live in the world of commerce where the object of production—the object of the entire society—is producing goods for the market to sell for profit. This is almost axiomatic; I don't think anybody could dispute it as a valid statement of the kind of world we live in. But it leads to certain things. It leads to pollution. It leads to unsafe vehicles. It leads to the sale of drugs that have not been tested. Profit is the important thing.

Things are not produced for the needs of mankind. Yet we owe a great debt of gratitude to the capitalist system. It played a very important role in social evolution and was a very necessary stage in social evolution. What it did was to socialize production, as Brother Platt pointed out. It increased by various degrees the speed of communication, the speed of transportation, and technology. It has created a world in which, today, man has solved the problem of production. We live in a period of potential overabundance. In other words, capitalism, as necessary as it was in social evolution, has passed its zenith; it has fulfilled its historic mission. It's now ready for a social change.

When I hear this talk of gradualism, as offered by Franklin Long in answer to an intelligent question by a young fellow, as though it's a slow long process necessary before you can have sanity, before you can have a world fit for human beings, it bespeaks very voluminously of not really understanding the world in which we live today. The gradual process has been completed. The gradualism necessary, the evolutionary changes, the transformation of handicraft production, hand tools, and the modern gigantic socialized machine are an accomplished fact now. With the gigantic socialized machine, with the tremendous development in technology, the world is now ripe to make this abundance available to all mankind.

I will tell you what is lacking. It is not the scientific inventions, not by a long shot. I'm not making derogatory remarks about intellectuals and scientists because I have a lot of respect for them; I think they are very important. Wald and Weisskopf pointed out what is lacking: It's men! Human beings are lacking. We underestimate the man in the street. We underestimate his capacity, his potentials. We look at him as though he were almost second class. When human beings become aware of the nature of society, when they have the understanding and the knowledge, there is no more powerful weapon than an idea come of age. Victor Hugo very soundly made the point that there is nothing stronger than an idea come of age. When an idea comes of age it's as strong as the strongest armies. That's the kind of period we live in now.

When you talk about science and the future of man, it's important to realize that now we are ready to live without money, without capital, without commerce, without waste slavery—a time for common ownership, for democratic control, where everything in and on the earth is the possession of all mankind. That is not a fantastic utopia; it's historical necessity.

Man today is faced with a dilemma—either survival or extinction. And he wants to survive. Man is not going to commit suicide. Necessity is a great drive, and the very necessity of our modern times is compelling people to wake up. The background of all this discontent today is almost an instinctive realization that this is 1969, practically 1970, and it's later than you think. We should·devote our efforts not to just trying to patch up the system with the New Deals and the Keynesians which have proved futile, absolutely futile, in abolishing poverty, abolishing wars, and abolishing the problems of society. We don't need these inventions. What we need is a determined vast majority of human beings to wake up and realize the necessity today is a world fit for human beings. It's possible and it's necessary!

Question (unidentified speaker): Professor Wald, do you think there is a need for a national advertising agency today to really push for good things like giving up cigarette smoking and similar things?

Professor Wald: Advertising is just another business. It's a very, very big business. And who would own it, who would control that national advertising agency? I thought that, in a way, we have that agency in the present national administration. (Applause and laughter.)

Question (unidentified speaker): Dr. Long, you spoke of the need for analytical thinkers. I wonder how you saw us staying out of the trap, as the machine now operates, of educating our graduate students in science where so much of our work is so highly specialized? How can we remedy this problem when you talk about the need for broad-based analytical thinkers?

Professor Long: The one thing that I am absolutely sure of is that in order to accomplish the kinds of changes within the university which are needed to respond in the ways that we have discussed, to do the things that John Platt has been talking about, it will require a great deal of very hard work. It's not going to be done by scientists working nights and weekends. It's going to be done by people who commit themselves, as Platt implies, genuinely full time to these problems. You only get the kind of cross-fertilization of ideas and approaches that these problems require by putting people together continuously. And it is into that milieu that we have to direct graduate students. We have to have graduate students involved in this broad, interdisciplinary approach as well as researchers. We need some new social organizations, structures in our universities, to permit this.

Let me just comment on one other point, which perhaps I didn't emphasize enough. I absolutely agree with George Wald and Lewis Mumford in saying that these things go well beyond all of the sciences, however defined. I was talking to a philosopher about some of these problems and how they might be solved, and he said that most of these problems, of the environment and urban society and poverty, fundamentally are problems of morals and morality. Of course, he is right. Without that spirit of what the problems are about, these multidisciplinary studies won't work.

Professor Weisskopf: I would like to add something on this last point. I fully agree with Franklin Long, but I think that he did not go far enough. I agree with our speaker that the way we educate our graduate students is wrong. We make them specialists in a small field of science. We must counteract this fully. But if we counteract this just by having them work in some of those projects that deal with the immediate necessities—which is certainly a legitimate activity—then that would not be enough. We must provide an education where people are educated in science as a whole, where they study nature, the whole of nature, for its own sake.

I mention this again, in spite of the fact that it is not fashionable, because it is necessary; because we can only get both the knowledge to solve the great practical problems of the day and the attitude necessary to care to solve these problems if we have people who are trained—and not only in the artisan sense—and deeply dedicated to nature and to the exploration of the secrets of nature. This is what we need.

John Platt says the scientists should do all this. Yes! Because they are the right people; because they are trained in that spirit I speak of. But all of our graduate students must not be moved in that direction. Basic science is only 5 percent of science and development. If we don't have that most gifted, dedicated 5 percent who really study nature for it's own sake, then we will be lost when ten years have past and we are in need of a new generation.

Professor Carovillano: Two points of view are being advanced. Professor Weisskopf says, yes, these problems must be worked on and at the same time we must not forsake basic research because it is the basis of our most fundamental knowledge and understanding of nature, and if we are going to get at these enormous problems that beset us, we must, of course, have a very broad base of fundamental knowledge. John Platt says that we don't have time to do certain things over long ranges because of the crises that beset us. Just because of the time scales, it is clear that we must work on these problems and we must work on them now. Now I ask, which of us are to continue to do the basic research and which of us are to give it up to respond to the pressing problems of society? It's difficult to see which part is for the greater good.

Professor Platt: Long-range science is useless unless we survive to use it.

Professor Wald: We'll have to manage both.

Professor Weisskopf: In this country, and I think in all countries, people always try to follow fashions. One day *this* is the thing to do, and the other *that* is the thing to do. However, mankind and the culture of mankind is what we have. It is here because there are many kinds of people. There should be and there must be many kinds of people. But this is really what I have to say: Graduate students—not only graduate students but everybody— should do what they think is most important. This is why I wish John Platt success, and I myself may help. Yet, we must leave room, freedom, and dignity for those people who want to start when they are young and go on to be basic scientists. If we create—as I think we are now doing—an atmosphere where, if a

young man goes to basic science then he is regarded as an ego-
tist, a social parasite, or worse, then I believe that we would lose
not only one of our most important tools but almost the basis of
our creed as George Wald presented it.

I am sure that the majority of our science students will not go
into basic science, and in fact they never have. But there will be
those who feel that basic science is a very great thing; people
who believe that science is important, indeed supremely important.
These are the right ones, the only ones to do basic science. But
you should not tell them that they are parasites of society. I fear
that the tendency is in this direction. That is what I want to prevent. (Applause.)

Comment (unidentified speaker): I would like to follow up on this
because I really don't agree with you. What I had intended to be a
question no longer is a question and will be offered partially as a
comment.

I don't think that the problem in the past has been that there was
no opportunity within the universities for a person to become a
scientist. Rather, the problem is that there was no room within
the universities with the traditional setup of science departments
for a person to direct himself into those areas referred to by Dr.
Platt. Science is now in this situation where it is under attack because people have been pushed into a mold, or have dropped out
altogether, and have not been able to move into society and
constructively help make decisions as to what directions society
should follow. I think that the university as it exists today is not
what Dr. Platt would have, but I don't think it's what you would have
either. You recognize that there is no such thing as chemistry,
physics, biology—it's just man and nature. But we don't have a
university that studies man and nature. We have a place that's bro-
ken up into these traditional disciplines. We don't have what Dr.
Platt desires and we don't have what you want. In what direction would you have us go to bring about your concept?

Professor Weisskopf: In both directions. In fact, I agree with you fully. I must not have expressed myself clearly because I was trying to make the same points, namely: that the present science establishment is not satisfactory, that it is a factory for creating specialized knowledge, and that we must go in both directions. I am very much in agreement with John Platt and with what he wants to do. My point is that his practical proposals need to be based upon a philosophy. Only then can they be accepted and realized. One part—only one part but an important one—of that philosophy is the profession of true basic science. I also agree with you that we have somewhat falsified—not all of us but many of us—basic science. So in answer, we must do both. Both are very necessary. It is true that to accomplish the practical applications we will need more people. But basic science also will need people, and we must create an atmosphere where we can develop what I call natural philosophy. That's the philosophy we need for a better future.

Professor Wald: I think what Victor Weisskopf is asking for is not only intelligence and learning and experiment but the committed man.

I'm going to ask John Platt a question. John Platt is an old friend, and I must say that when he presents this situation I am full of enthusiasm for his clarity and his contribution. But when he presents the solution in the form I heard the other day, with 220 task forces or committees each going to work on some subheading, I feel a profound sense of depression. I'm going to ask him a question and I guess he can answer this easily. So John, name me a social invention that has come out of committee.

Professor Platt: The Test Ban Treaty, the Hot Line, the Non-Proliferation Treaty, the Antarctic Treaty, were all generated by the committees that are called Pugwash.

Professor Wald: I'll have to look into those. (Laughter.)

Professor Long: In fairness to John Platt, it must be noted that the absolute essence of his proposal is not for committees. It is for groups of full-time workers totally committed to some of these problems. George Wald is right; if we depend upon faculty committees that meet monthly to go over things and exchange views, we will get one thing. In order to get what John Platt wants, however, it's full-time commitment—people, money, students—that is required. It's just a different game, and I don't think we should minimize the difference.

Comment (Professor I. Asimov, Boston University): My speciality is getting all confused in semantics. Now I hear talk that makes it seem as though basic research and the kind of applied research that Dr. Platt is speaking about are somehow two different things which don't overlap. If you do one, you can't do the other. Now, as a matter of fact, this idea exists only in men's minds. Basic research is never so pure and virginal that it really has no inspiration in the actual problems of the day. For example, a great deal of basic research in heat and thermodynamics was inspired by the dirty old steam engine. Now, I think that what we need is basic research in saving society. This means that by no means should we establish some sort of research program to save our society because that's applied. Let's, instead, set up what we might call generalized societies and figure out how to save them. Then the applied scientists can merely stick our society in where x is and can save ours. That way, we'll all be happy. (Applause.)

Comment (unidentified): On the issue of saving society, about thirty or forty years ago a group called Technocracy came up with what some people have considered to be a method of resolving problems of society. Basically, they felt that economics had to be made into a physical science. This was necessary because once society had arrived at a certain point of abundance (which they felt was a fact at that time), the only thing that could be done with the abundance was to distribute it—the abundance of goods, of food, of housing, and the abundance of knowledge. Of course,

this hasn't been done yet. Their contention is that all economic institutions, all concepts of money, all financial institutions, all
tax structures are obsolete. Technology has made them obsolete
by producing abundance which destroys the value basis of money, which scarcity determines. They felt that by restating economics on a physical basis—as energy, which is the common de-
nominator of all goods and services—we then could control our
technology. They advocated simply that we apply our scientific
knowledge without the restrictions of politics and money. I think
that, in the end, this is what will have to be advocated or else
we won't survive.

Professor Carovillano: Mr. Mumford, we have all grown impatient
to hear from you again.

Professor Lewis Mumford: The real question is whether scientific
research—in its broadest and fullest scope, which we all honor—
will be sufficient. Will that sufficiently change our minds, our own
way of living, or will we still be in this rut?

We are not attached to the capitalist system or to the Pentagon.
We are attached to a power system. You might call it the *pentagon
of power.* Its ingredients are power in the sense of physical energy, power as productivity, power as profit, power as publicity,
which is an essential part of this whole scheme of life, and power
with pleasure as the ultimate end of existence whether offered by
narcotics and drugs or some other form of pleasurable excitement.

These are almost interchangeable parts of the same system. But
we can't change one part very easily at the bottom of it. The reason
it works is that *profit* plays something of the part that the pleasure
center seems to do in the investigation of the reaction of animals
like monkeys. If you locate the pleasure center in the monkey's
brain, attach an electrode to it and have the amount of current
regulated by the monkey himself, the pleasure is so intense that
the monkey will be transfixed in this particular situation and will

disregard food and any other part of his life until, if carried far enough, he would die.

The profit center in our civilization plays the part that the pleasure center seems to do in the organism. We keep on pressing the controls in order to die of the pleasure from having profits that we are no longer able to consume, or use, or dispose of. This is a central problem. I feel it's so central that I spent a large part of my book [entitled, *The Pentagon of Power*] analyzing it and showing its origin and extent.

Opposite to the power system is an organic system that has built-in controls from the very start. No organism disposes of unlimited quantities of energy. All organisms have a way of regulating the amount of energy that can be acceptable to sustain life. The purposes of reactions of all organisms always have well-defined ends and well-defined limits. Nobody wants to grow to be ten feet tall, and Nature sees, on the whole, that there are limits, both high and low, to the size of an adequate human being. Aristotle pointed out that this also applied to ships. If a ship is too small, it can not carry cargo; if it is too big, it can not be maneuvered. These built-in organic controls are lacking in our power system.

The original mechanical world picture was magnificently developed in the seventeenth century, particularly by Galileo and those who followed him—Descartes and Newton. But this so-called objective world is just one of the creations of the human personality which the human culture has sustained. Until we understand the limitations of the human organism and understand that we must replace the mechanical world picture with an organic world picture, in which the human personality plays a central part, we won't really get to the bottom of our problem.

[To the session chairman:] Is that bad enough? (Applause.)

Professor Platt: I like Mr. Mumford's analogy of our society and its feedback stabilization system and the ways it becomes locked in on particular pursuits. The analogy was made to organic creatures and animals, noting that the animal has a design built into him. This is correct. But that design is built in by millions of years of evolution. Built in by survival. We haven't got time to wait for the survival of a few million civilizations like ours to find out which ones will develop and accumulate the feedbacks that promote survival. I think we have to *design* these feedbacks, these lock-ins, these pleasure principles, for our own society by the application of anticipatory intelligence. We cannot wait for this destruction that is upon us.

This is not an impossible task. The United States Constitution was a deliberate, rational design of feedback—in those days called checks and balances—attempting to balance the forces in society so as to provide a maximum of freedom with a maximum of effective governmental solution of large national problems. Within five years, the result was that the United States was transformed from a group of almost warring states—that couldn't raise money, that taxed each other, that were in continual rebellion—to a prosperous, growing economy.

We need the same kind of intelligent design of our internal feedbacks at this stage. It is like stepping up to a new stage of evolution. We now require a transition to a new hierarchical structure which has to be on a world scale instead of on a national scale. The brains required to assemble this hierarchical structure and to design the social feedbacks to give us freedom and humanism and the prosperity we all want, including the Third World, have to be developed in the next few years if we are not going to kill ourselves. (Applause.)

Comment (Professor Robert Faulkner, Boston College): I would like to take up a rather theoretical question which tunes into the question that Mr. Mumford raised: whether, in fact, further pursuit of scientific knowledge does have a sufficient answer. We have talked of the problems raised by modern science remarkably candidly, and my question concerns the suitability of science in providing the answers. Mr. Platt seeks the answers in a succession of social inventions, as applying the notion of invention so popular in technology to that of society. Every panel speaker sooner or later gets down to the supposition that from applied science our answers will come. I was disappointed to hear even from Mr. Mumford that the application of the objective technique which has been successful in the external world to our subjective world would, in fact, yield a solution. I wonder if it is true.

Professor Weisskopf: He didn't say that at all.

Professor Faulkner: If I may explain, I wonder whether the very method of science might be a source of some of our profound difficulties. Observe the adjectives that we use to describe it: observational, detached, objective, analytical. Each of these is a cool adjective, detached from human purposes, not oriented to human desires, pleasures, or excellence. Moreover, somehow the scientific method seems not to be too open to the things that experience or experiment cannot reveal. For example, beauty. Can the beauty of something come to be realized by an experiment upon it? Very unlikely. One wonders then whether precisely our acceptance of the limitations of the scientific method has much to do with the disintegration now present in our minds, or as Mr. Mumford so eloquently put it, the demoralization and disintegration of our civilization. There seems to come a question as to whether any values like music, or goodness, or excellence, have any rational basis.

Speaking out theoretically, I wonder if there is not a problem if practically we organized two hundred or even two thousand task forces on principal assignments, whether this would not, in fact, simply press 1984 closer and closer upon us under the new principle of planning or confining our attention to great problems. This is not to say that problems aren't to be studied. The question is whether the mode to study is to be that proposed by every speaker here. The question then is whether we need and have to take one more great leap into the future under the aegis of science, or whether turning back to some older way of *thinking*—I emphasize *thinking*—is not perhaps a more promising direction.

Comment (unidentified speaker): I am a dental medical student, and I thought I might hazard a guess as to why there are not too many students here. For a lot of young students today, the whole idea about talking about science and the future is, you might say, almost irrelevant to their personal experiences. There is another symposium going on in which people are analyzing essentially why scientists got us into the mess we are in now. One may or may not agree with that outlook. Many people here may say it's the political institutions that are at fault, or it's the fact that we haven't controlled our objectives and the irrational tides that have gotten us into this chaos. Last night I was speaking with the brother of a friend of mine who has transferred to New York. He has just come back from a liberal Midwestern college, and I asked him how it was out there. "Well," he said, "it's ridiculous, it's like a ball game. There's nothing going on there that's very interesting. There's really nothing real." He said that next term a Zen master was coming from Japan and he was to start meditations.

I think another problem that hasn't been raised is the whole drug situation. Why are young people turning to drugs? Why marijuana? Why LSD? Why hallucinatory drugs? As was touched on by Dr. Wald and by Dr. Mumford, I think that for a lot of young people, the future as we are talking about it here really doesn't exist. Especially if it is to be determined by science. People are trying to develop new ways of feeling and thinking about the world which maybe their parents did not communicate to them. I personally sense that a feeling of despair and depression has developed among an awful lot of people within the last year and a half.

It seems to me that there are two alternatives to a number of young people: either despair and depression and taking more drugs and disassociating from society, or a more apocalyptic solution. I think a lot of young people are talking about that. Also, a lot of young people are really antiscientific and a number are really anti-intellectual. They really do believe that science is perhaps an indecent word or evil. I do not think that personally, but a lot of people do. If the future of the world depends on technology getting us out of the mess that technology has created, and if an awful lot of bright young people don't even want any part of the technology, what kind of a predicament are we in then? Maybe the generation that got us in will have to get us out. Would anyone like to comment on that? (Applause.)

Professor Wald: I would like to comment on the course that Lewis Mumford has gone. He just told us he got the Leonardo Da Vinci award. I am getting another award ready for him. It's the "Angry Old Man Award." He's my favorite angry old man. There are others.

I would like to answer the last speaker in a way—in a way. You see, all men seek answers to the same questions: Whence we come. What kind of thing we are. And at least a hint of what might become of us. In trying to answer these questions, in every time and place, in every generation, men seek all kinds of paths. Zen has one of those paths. Science properly pursued is one of those paths.

We have a tradition. For me nowadays, I read the Prophets—in both Testaments. They are always saying the same thing—always the same thing. God isn't interested in your observances, your loud singing, your loud and ostentatious praying, your sacrifices. He wants you to be decent. Clothe the poor, feed the hungry, heal the sick! It was Paul who said those beautiful words: These three things abide; faith, hope, and love. I think the generation gap is a gap between a younger generation, the best of whom are looking for that faith, hope, and love and are insisting on finding them, and an older generation which has never found them and has gotten used to doing without them.

So that's my formula. Science takes you a long way, and for me the obligato in that journey is in the tradition. It's our Western tradition. Try the Prophets.

Professor Carovillano: At this point it will be necessary for us to adjourn. I would like to thank the panelists for their participation and the audience for its attendance.

Part III
Confrontation
Chairman: W. Seavey Joyce, S.J.

Chairman:
Robert Drinan, S.J., is best known as the dean of the Boston College Law School and for his active role as critic and public spokesman on many current issues of great importance. He is characterized by taking positive positions, highly unpopular ones when necessary, and has been most concerned with family law and church-state relationships in modern society. Father Drinan has spoken boldly in suggesting that because of the great diversity of opinions and attitudes the state should withdraw its controls on abortion and let this issue be treated on moral rather than political grounds. He has been to South Viet Nam investigating political conditions and prisoners and has expressed his convictions that the war supports a nondemocratic and repressive regime. Father Drinan has recently been made vice president and provost at Boston College but has promptly withdrawn from all administrative university positions in order to pursue his fortunes as a Democratic candidate for Congress.

Science and the Revolution in the Third World
Robert F. Drinan, S.J.

Whenever I talk to or think about distinguished scientists I feel like those people in the fourteenth and fifteenth centuries who, after printing was invented, still could not read. I feel illiterate. And I am afraid that many, many nonscientists feel that way.

All of these individuals feel to some extent locked out of today's society. I think it is fair to say that science in this day and age—and I am talking about science and not the scientist—has become apotheosized. We have become so addicted to new knowledge that any modification, any thrust, suggesting that perhaps we have to set different priorities immediately comes out with a negative tone.

The test between religion and science of the last century is over.

But the same onslaught of empirical knowledge which caused religions to shrink back from science in the last century has caused different people to shrink back from science in this particular day and age. Call these persons humanists if you will, or people that believe in humanity or brotherhood, but there is a new war on science which is emerging. That war was evident yesterday in the pickets and demonstrations against this distinguished gathering. Those who wage this war do not say that we are against science, but simply that we want a new order of humanity. The priorities that now exist in science are offensive to that new world which the young and others are seeking to create. This is a war against the use of science for the creation of superfluities, for militaristic purposes, for the purpose of building the ABM or the MIRV or even for the development of conventional arms.

I suggest to you that in the next decade we will have a massive dialogue or confrontation with regard to the priorities which we will assign to science. I think it is fair to say that the government and private foundations have fallen into the contemporary apotheosis of empirical science. The National Science Foundation has given millions of dollars to the development of the physical sciences. But it was only after a long struggle that the National Humanities Foundation was established with very, very small endowments compared to the National Science Foundation. Similarly, the other foundations have gone along with this tidal wave of empirical research in the natural sciences. For example, as of two years ago the Ford Foundation had given vast sums to a field such as education in chemical engineering compared to relatively puny sums given to legal education (a clear distortion in my priorities!).

To be sure everyone wants more answers from science. We want to know, for example, why of the hundred million people born into the entire world this year, 4 percent will have grave physical or mental defects at the time of birth. The latest national concern is to stop the pollution of the environment. But can we have all these things and these priorities in the Atlantic world—the world of

Europe and America? Can we have this consistently with the voice of our conscience, when one-half to two-thirds of humanity in the underdeveloped countries—some two billion people—do not even share in the rudimentary things that science has created for us in the past century?

To some extent science is like the church. Both have intellectual and spiritual riches. One of the eternal questions for the Christian churches is this: Shall we concern ourselves only with the one-sixth of humanity which is Christian or shall we go on to new frontiers and carry the faith, to develop a culture saturated with religious principles that will be a beacon light throughout the world? Science has that same problem: Shall we go and transform other societies into something like ourselves, or shall we stay here and do pure research, the research that will be effective in the twenty-first century?

Science sees the breakthroughs ahead, and I am not faulting the scientist. If I were in that area, I would probably be intrigued by and addicted to the mysteries of the DNA, to the capacity of prolonging life, and all of these things that we do not fully understand. Scientists are like theologians. They are like missionaries. They decide what is the more important thing, the exploration of their rarified area of knowledge or the creation of a more human world. The scientists go to the moon and, I suppose, they will go to Mars in the next decade.

But I suggest that in the area of the Third World we must ask ourselves basic, fundamental questions. We simply have to have some type of reordering of priorities. The Vatican Council put it this way: All of humanity must strive to develop a world which is a dwelling worthy of the whole human family; man should subdue the earth by bringing creation to perfection, to elevate the whole human family into a situation where every man can have human dignity. Or put it another way: Can we insist that no public authority in the United States or in Western Europe may divert science to serve its own practical or economical or political interests? I think that it is rather clear, as Pope John suggested in *Pacem in Terris,* that political and economic interests have captured the creativity of scientists.

I suggest, therefore, three propositions which I will develop. First, the revolutions in the Third World are inevitable and are inexorable. They will not brook delay. Second, science is the word that will be hated and is already hated in the Third World as representative of a force which is depriving the peoples of the Third World of the necessities of life. Third, I will suggest some affirmative ways to alter America's priorities without, I hope, suggesting or implying that science must slow down in its basic research in the highly developed countries.

The revolution in the Third World it seems to me is caught in the famous words of President Kennedy: "Those who make peaceful revolution impossible make violent revolution inevitable." In a famous letter written by seventeen Roman Catholic bishops from the Third World these prelates, after living among these benighted people who look at television, who hear their wireless and see a world where human dignity is attainable, state that it is the duty of all citizens and of all nations to allow these people to acquire human decency and dignity. The seventeen bishops gave the familiar statement that it is not wrong for the hungry man to steal bread from the affluent; they suggested that words are not enough on the part of the highly developed nations. Clearly and forcefully they suggested that direct and militant actions might be allowable for the citizens of those underdeveloped nations who see their brother nations living in affluence.

Words are no longer enough on our part. The United Nations Conference on Trade and Development some years ago brought together eighty of the poorer nations as an international lobby for the usual things, such as steadier commodity prices and more favorable tariff structures. But that organization has not been able to move simply because they cannot confront the powers of the first and second world. America has made a commitment to these nations, at least in the famous words of President Kennedy in his Inaugural Address. In talking about the Third World President Kennedy said:

"To those people in the huts and villages of half the globe struggling to break the bonds of mass misery, we pledge our best efforts to help them to help themselves for whatever period is required. Not because the Communists may be doing it, not because we seek their votes, but because it is right. If a free society cannot help the many who are poor it cannot save the few who are rich."

The substance of these remarks goes way back to the Truman Doctrine, to January 20, 1949, when President Truman issued the famous Four Points. The first three were forgotten almost immediately. But point four indicated that somehow we will give aid or restitution or reparation or indemnification to the underdeveloped countries. Thus, foreign aid was born. Shortly thereafter, foreign aid, however, was used as a weapon in the cold war and to some extent was diverted to help the economic, political, and military objectives of the United States.

During the last twenty-five years of the atomic or nuclear period, the United States has made at least a token commitment to help the tribesmen in Africa and the dispossessed in Latin America. Perhaps this commitment is the result of our Judeo-Christian tradition. Perhaps we remember the words of Saint James who said in his epistle: "If a brother or sister be naked, if they lack their daily nourishment, and one of you says to them, go in peace, be warmed, and be filled, without giving them what is necessary, is not that unjust?" The same thing was said in 1967, by Pope Paul in that famous encyclical on the development of nations which shook so many people—particularly the "establishment" in the underdeveloped countries. Pope Paul said, "We must strive to build a world where every man, no matter what his race, religion, or nationality, can live a fully human life, can be free from servitude imposed upon him by other men or by natural forces over which he has not sufficient control."

I suggest to you that the Third World *will* demand justice sooner rather than later. How? Can the nations of the Third World acquire an intercontinental ballistic missile and threaten to bomb the highly developed nations? I think this is quite likely. If one reviews conditions that precipitated the mobs that brought about the French Revolution or if one reflects on the anger and indignation of the oppressed serfs that brought about the Russian Revolution, I think it is clear to say that one-half of humanity that is in the underdeveloped countries can organize in ways that we would not even think possible at this time.

What will be their cry? Perhaps they will say that they want reparations for the wealth that was stolen from them by the colonial powers. I think myself that would be a very justifiable plea.

To what extent will their plea be violent? However one looks at it, it seems that there will be a conflagration unless we, the Western world, somehow recognize that it is impossible to continue in the present situation, where America, with 6 percent of the entire population of the earth, controls 60 to 70 percent of all of its wealth!

I plead with scientists to recognize the urgent message in Barbara Ward's many writings on the rich and poor nations—her burning message is that regardless of what we may do by way of tokenism in the Third World the situation as it now obtains there cannot continue.

Will there be an alliance between the 22 million black Americans and some 200 million black Africans? I suggest that is very possible, just as the Black Muslims now have an alliance and have joined, at least in spirit, with the Al Fatah to advance Arab power—in this instance, against Israel. The anger, the indignation, the voicelessness, and the powerlessness of the Third World cannot do anything except explode.

I come, therefore, to my second point: What can science do about this? Science will become, if it has not already become, the most hated word in the Third World. The peoples in these underdeveloped nations look and see that the skilled technocrats have all been preempted by the industrialized world. Engineers and scientists do not come to their forgotten countries. Somehow solutions do not filter down to them.

If I may be personal, I would like to relate two instances that impressed this upon me. About two years ago in the Belgian Congo, a high-placed American official had fourteen "ugly Americans" for dinner one night. We were attending an international meeting of a federation of universities in Kinshasa. The American official was lamenting the fact that his foreign aid had been cut that year by a substantial sum and that as a result some children in the Congo were not receiving vaccine for smallpox and other diseases. Children were dying that night while we ate our imported steak. All during that day, two hundred Congolese men had marched around the office of American officials in the Congo. Can we allow this situation to continue? What can we expect from the Third World as foreign aid is trimmed, now to its lowest level since the nuclear age began?

On another occasion, last year in Saigon, I visited a fantastic medical laboratory, established there with the aid of AID and American foundations, where scientists were researching into cures for the twenty-first century. Here these men pursued rarefied research in enzymes while outside the door of this magnificent clinic and laboratory, outside the very door, people were dying of diseases of the nineteenth century!

This imbalance, this disproportion between the great wealth and the great poverty will be blamed on science. Now I would be the first to say that it is unfair to blame scientists. (I usually blame lawyers, since I talk to them more than anybody else and feel that they should be our moral architects!) Nevertheless, in this instance at least, the scientists are the most influential segment of society. They are the sought-after people. They are the needed ones, the esteemed ones. They are a whole group within a group. And what do they do? According to the rhetoric of the "kids," they get corrupted. They get preempted. They sell out to the Establishment.

I suppose, in the ideal order, we should really be pursuing pure research and applied science. But I worry more and more, and I hope that scientists worry more and more, about the one-half of humanity which has virtually none of the very primitive fundamentals that have come to us from science and technocracy.

Conceive of the moral power which scientists have! In 1955, fifty-two Nobel laureates wrote a famous statement which asserted "that all nations must come to a decision to renounce force as a final resort of policy." And two years later, in the well-known bomb test appeal signed by more than eleven thousand scientists from forty-nine nations all around the world, the signatories asked for a stop to nuclear testing. In their statement and effort I think that you can trace the ultimate causes of the Nuclear Test Ban Treaty of 1963.

Is it a good thing for scientists to get mixed up in politics? I suggest that it is really not a question of politics, but one of basic morality. Scientists cannot abdicate their humanity to their science. Scientists should not regard themselves as just "hired hands" to think great thoughts and to explore the universe. It is unrealistic to think that we can continue with a dehumanized science.

Imagine the power of the ten thousand scientists at this particular AAAS convention. I am not faulting the president of this organization, but I read in the newspaper today what he said on the space program. I am sure that the students would be the first ones to point out that he is the president of the University Corporation for Atmospheric Research at Boulder, Colorado, and consequently has a slight interest in the further lunar explorations which he is advocating. He suggests that we use space stations as monitoring devices. I am not necessarily against that, but I look for a new vision which will allow us somehow to deploy all of the scientific knowledge that we have for the enlargement of humanity. Dr. Roberts, to be sure, laments and deplores the enlargement of the arsenals of the United States and of Russia by reason of space technology. But now and in the decade to come scientists must have a basic and growing conviction that they are more powerful in this area than anyone else.

My third point is to make some recommendations. The famous essay of C.P. Snow, *The Two Cultures and the Scientific Revolution,** contrasted those in the scientific world and those in the humanistic world and pointed out that there was no common language, no common discourse between these two groups. An analogy can be made in that we in the industrialized world cannot really speak to the underdeveloped world; there are two worlds here that are growing apart in a more and more frightening way. Science, after all, has emerged and proceeds in the highly industrialized, tiny area of the world that we call the West. Scientists have to stretch their minds in new directions. Can capitalism, with democracy as we conceive it, build a just world? Can the United States continue to furnish arms to every country that buys or wants them? And what about disarmament? These may sound like radical issues, but I am suggesting that science cannot abdicate its fundamental responsibilities in these areas.

* C.P. Snow, *The Two Cultures and the Scientific Revolution* (New York: Cambridge University Press, 1959).

Alfred Nobel, the founder of the Nobel Peace Prize, wished seventy-five years ago for the invention of a "substance or a machine with such terrible power of mass destruction that war would thereby be made impossible forever." Scientists discovered that machine a few years ago. They discovered fissionable substances, uranium and plutonium, which contain an explosive energy 10 million times more than that of Nobel's favorite explosive, nitroglycerin. Consequently, scientists have changed, *radically changed,* the entire world. It is impossible for them to say now that others must solve the dilemmas that have evolved.

The fundamental dilemma, or at least one of the dilemmas, is the whole question of capitalism operating within a democracy. What is the alternative? I am not going to dream here about a world government with the *one* army that could wage war, although this may well be the only alternative. As I understand the rhetoric of the young and the questions of some middle-aged persons, they say that unregulated capitalism cannot be expected voluntarily to alter its course in order to meet the revolutions in the Third World.

I know that this is a vast, vast subject, but I think it is an essential one, a crucial one. Perhaps it can be understood by noting the virtual impossibility of any democratically elected or appointed official in America altering in any fundamental way the foreign policy of this country. For better or for worse, for twenty years our foreign policy has been a bipartisan policy of containment, with the alleged enemy being communism. But communism has many faces, as we know, and that policy, in my judgment, is now outmoded. But can we expect any person who is responsive to an electorate, who has been appointed by some Democrat or Republican, to come out in bold terms and say that we must reorder our thinking and that all of the anticommunist fundamentals of our foreign policy must be modified? I suggest that only the scientists, with their enormous moral power, can raise this fundamental question. I suggest moreover that it is the first question that has to be raised.

Next, can the United States continue as the largest supplier of weapons to the nations of the world? Books continue to come out on this issue, and they horrify us. Perhaps the most recent one is George Thayer's book, *The War Business: The International Trade in Armaments* (1969).* He has demonstrated, as never before, that in the last analysis the United States is responsible for allowing the introduction of unneeded, unmanageable, sophisticated weapons systems into the underdeveloped countries. John Kenneth Galbraith revealed in his book *How to Control the Military*+ that there are seven hundred retired generals, admirals, and navy captains employed by the ten largest defense contractors—thus intensifying the linkage between the manufacturers of arms, the Pentagon itself, and the Third World. I know this is a large subject for scientists to take on, but I think that they, and perhaps they alone, have the prestige to deal with it effectively.

My third point is unilateral disarmament. Everybody wants bilateral disarmament, but is that really feasible? Would it be possible for more scientists to support some form of unilateral disarmament, as Erich Fromm suggested way back in 1961? I know that this is not a popular idea, but in 1960 in the *Bulletin of the Atomic Scientists,*‡ scientist Charles Osgood suggested graduated unilateral action or disengagement. And yet that idea has somehow faded away during the past decade so that disarmament now is tired. One reads about the SALT talks in Helsinki and Vienna without really any hope, and people feel that MIRV is going to happen.

---

* George Thayer, *The War Business: The International Trade in Armaments* (New York: Simon and Schuster, 1969).
+ John K. Galbraith, *How to Control the Military* (Garden City, N.Y.: Doubleday & Co., 1969).
‡ Charles Osgood, "The Case for Graduated Unilateral Disengagement," *Bulletin of the Atomic Scientists* 16(1960): 127-131.

All of these frustrations are alienating and angering the young and, I am sure, bothering the scientists enormously. Scientists cannot evade their responsibilities. For persons of their highly sophisticated caliber—really for everyone in the United States—there is a challenge which is new. We have never had this before. For almost twenty-five years, this terrible rivalry in superfluous strategic arms between Russia and the United States has taken on the name of a game. All we know is that every time the game is played there are a few billion dollars more spent on hardware and the name of science and the name of Western culture continues to decline in the estimation of the Third World.

My final point is whether the United States has some role in the entire world. This is a matter on which, frankly, I think we are somewhat schizophrenic. We go back and forth. We used to be fundamentally isolationists. Now many people who are opposed to Viet Nam tend to be neo-isolationists, saying, in effect, "Let us live our lives and let the rest of the world find their own way."

I suggest to you that we cannot do that because built right into our mystique—the mystique of the American people—is the idea that somehow we are the purveyors of good news to the entire world. I could cite Wilson's Fourteen Points or Franklin Roosevelt's Four Freedoms, but it is better expressed in the older words of Abraham Lincoln:

"I have often inquired of myself what great principle or idea it was that kept this confederacy so long together. It was not the mere matter of the separation of the colonies from the motherland, but something in that Declaration giving liberty not alone to the people of this country but hope to the world for all future time. It was that which gave promise that, in due time, the weight should be lifted from the shoulders of all men and that all should have an equal chance."

This is the sentiment embodied in the Declaration of Independence.

Science in the world after Viet Nam has to face up to that dream that is embodied in our American mystique. As we look out and see that almost two-thirds of humanity does not enjoy the affluence of the richest nation in the history of the earth, we have to agree that countless ways must be devised so that they, too, will share what we have in equality and human dignity. I think that scientists have before them an opportunity of surpassing importance and urgency. I think it is the most important opportunity which scientists have had in the history of the Western world. If they can refine and define this moral dilemma of agonizing implications, if they can go back to the religious and political traditions and articulate them in sociopolitical ways, then they may be able to persuade this nation that it has a moral commitment to mankind, and that we should carry out that commitment, not because it is useful to America but because it is right.

Chairman:
The significance of this symposium is greatly enhanced by the participation of Senator Muskie. As one of the most concerned and most knowledgeable of our present day legislators, he has developed an impressive record in many important areas of national concern. Before the present national and worldwide preoccupation with the deteriorating environment, Senator Muskie was holding hearings and investigating the magnitude of the problem. If as he says, "the wind is at our back" regarding the popular awareness of global pollution and the clamor for a quality environment, such a favorable climate has been in no small measure generated by the understanding of the problem that he and his co-workers have stimulated.

Crisis of Man and His Environment
Edmund S. Muskie

You know, I spent the Christmas holidays in Washington with my family although we had planned to spend them in Maine. We changed our plans because the Congress prolonged its labors to the point where it seemed unwise and impractical to try to work out the logistics of moving five children and a family to Maine on short notice. And so we enjoyed the holidays in Washington reading about your difficulties in New England with the weather. I must say it was somewhat humiliating to note that northern New England seemed suddenly to have developed an incapacity for dealing with snow. Things have changed since I was a boy. And so it is appropriate, I think, to be talking about the environment in these kinds of New England circumstances.

I can't resist beginning with a Maine story that has some relevance to our frustrations and our confusion about this subject. In most of our stories about out-of-staters, we like to pull their leg. This one also concerns an out-of-stater, who is lost in Maine and stopped to ask directions of a native. He said, "Will this road go to Millinocket?" The native's reply said, "No, it don't go nowhere, it's been here all my life." "Well, if I should turn around, will it take me back where I came from?" "Well, I don't know, I don't know where you came from." Well, at this point the out-of-stater was somewhat upset and he said, "You know, I think you are the dumbest man I have ever met." The native said, "That may be so, but I ain't lost."

Eighty-six years ago Henry Augustus Roland told the American Association for the Advancement of Science that American science is a thing of the future and not of the present or past. Today we may well ask whether science, which is responsible for much of our past and our present, has prevented the future, or whether it will make possible a worthwhile future for man? Until very recently the question of man's future and its vulnerability was related to the threat of nuclear war. That threat remains. But it has been joined by the threat of environmental contamination. Man has so misused the fruits of scientific endeavor, he threatens his own existence. Some threats come in bits and pieces, the by-products of our industrial economy.

The daily newspapers carry samples of such threats as in the following two examples published during the weekend before Christmas. Here is one. When the sulfur dioxide content of the air in New York City rises above two-tenths of a part per million, ten to twenty people die as a result. In the past five years sulfur dioxide has reached this level at least once every ten days. Another, the modern industrial economy is dependent upon hazardous materials that are shipped throughout the country. In the last five years over fifty cities and towns have had to be evacuated as a result of accidents involving hazardous materials.

Some threats come from defense projects designed to protect our national security, as noted in the following item in the press. The president's statement on germ warfare has not ruled out the production of toxins. The Department of Defense does not find in the president's directive any specific prohibitions to the production of toxins. Some threats are the result of efforts to dispose of wastes from the conversion of materials and energy. Outside Denver, for example, a farmer's well produces the weed killer 24D. His neighbor's well flows gasoline. In Ponca City, Oklahoma, springs bubble refined motor oil into residential basements. The culprit in all of these cases is called deep well disposal. Under this system, billions of gallons of salt water mixed with oil and other liquid wastes are being pumped back into the ground as a way of disposing of them. Texas alone has thirty thousand such wells.

On the East Coast, one scientist has proposed the construction of a forty-eight-inch eighty-mile-long pipeline to carry municipal and industrial wastes from the lower Delaware River basin out into the Atlantic Ocean to discharge up to 4 million gallons a day well beyond the continental shelf. This scheme would reduce pollution in the Delaware basin at the expense of the ocean. It is like so many other examples of pollution control programs. It proposes to dump the load on someone else downstream, but we are learning that there is no one else downstream from us. We have made the world smaller with our population increases and our transportation advances. We have contaminated the land, the water we use, and the air we breathe with wastes of our own making. We have gone beyond the point where the issue of conservation is limited to those who want to protect a stream or a forest or a stretch of shoreline. That protection is still needed but it is not the central issue.

The central issue is the health of man wherever he lives and whatever his station. This is the issue the young people understand. This is the issue which has placed the environment at the center of campus concerns following the Viet Nam war. This is the issue which cuts across economic, social, and racial lines. It binds the suburbs to the cities in a common life-related problem.

Such an issue, touching as it does the lives of the young and the old, the rich and the poor, is a deep and strong political issue. It is real and therefore susceptible to emotional appeals. It is broad and therefore subject to many uses. When such an issue arises, would-be leaders and voters will look for scapegoats; and those who resist change, especially in their own behavior patterns, will dismiss environmental complaints as uninformed demagoguery.

Scapegoats will not be hard to find. There are industrial and business leaders who reject any responsibility for pollution or its cleanup. There are public officials who avoid the unpleasant encounters so necessary to change. There are managers of public programs, civil and military, too mission oriented to admit any responsibility for protecting the environment. And there are scientists whose commitment to their own projects has been so total they have ignored the environmental consequences of their work. Protection of man and his environment cannot be achieved by casting out scapegoats, however much they deserve it. Neither can be advanced by scorn for environmental complaints.

The pollution of our environment is not the product of a small band of men and it is not the product of our particular economic system. It is the special product of any society which places the consumption of goods and services high on its scale of values and which has the means to provide those goods and services in abundance. It is not who owns the means of production but how the means of production are managed that determines the impact of the industrial-technological society on the environment. Pollution is a worldwide problem which will not give way to political code words. It can be exorcised only through intelligent public action based on an understanding of its causes and appreciation of its constantly changing aspects and a comprehension of its implications.

Now I may seem to have said that nothing has been done to protect man from the follies of his own wastes. Since the latter part of the 1940s, we have been chipping away at the obvious sources of water pollution from municipal and industrial sources. The greater part of our limited success in this area has come in the last six years with the development of our water quality standards program, a substantial increase in our commitment to build waste-treatment plants, and attacks on specific problems such as oil pollution, thermal pollution, and industrial pollution. But let us concede that up to this point, most of this limited success has been related to the development of public policy and not its implementation on the scale that is required. Work on the problems of air pollution came later because the threats were not so obvious and because we did not make the connection between it and public health. Nevertheless, we have launched a program that is designed to achieve high standards of air quality in all parts of the nation. It is a program that deals with moving and stationary sources, and most important, it is organized to build on scientific data concerning the effects of pollutants upon human welfare and to stimulate the gathering and use of such data as it relates to public health. Finally, we are in the process of converting the solid waste control program from an exercise in the disposal of waste to an attempt to reduce the volume of waste in our society and to encourage the more efficient use and reuse of materials and energy.

These are programs dealing with the obvious and straightforward problems of pollution, the physical by-products of our activities and our production. They are, if you will, the first stage problems of an environmental protection program. The next stage, which will be tougher, will involve the organization of our public institutions to deal with the more subtle and pervasive questions of land and resource use, of population distribution, and industrial location of hazardous substances and their release into the environment; of noise and aesthetic pollution; of private decisions that result in the unleashing of pollution generating technology; of ecological balance and urban design.

Such questions affect the way we organize federal, state, and local government and the relevance, indeed, of the federal system; the way we organize planning decisions, systems of taxation, public works projects, support of research and development and even decisions on defense policies. Increasingly, such questions will involve our relations with other countries in this hemisphere, in Europe, Africa, and Asia. The atmosphere of this planet, after all, is indivisible. Efforts are under way to provide a base for this second stage effort in the federal government. The Congress, for example, sent to the president the National Environmental Policy Act of 1969. That act presents a statement of national policy on the environment. It directs all federal agencies to comply with that policy. It provides the president with a council on environmental quality and it requires the president to submit to the Congress and to the people an annual environmental quality report.

The Water Quality Improvement Act of 1969, now in conference between the Senate and the House, complements that act in two important respects. First, it expands the requirements for federal compliance with water quality standards to include activities and projects supported or authorized by the federal government. Second, it establishes an Office of Environmental Quality to provide staff support to the president and the Environmental Quality Council. Those of you who are sensitive to the implications of words for politicians, and I must admit that we belabor them in a way that dulls those sensitivities, have probably noted our shift from pollution control and abatement to environmental quality. This is an important shift. The object of our proposals for expanding the president's capacity to deal with federal responsibilities in environmental protection is partially a recognition of the increasing complexity of the problem and partially an admission of the confusion which has governed our past efforts.

We face a similar organizational problem in the Congress. Several members of the Senate and the House are moving to create a nonlegislative joint committee on environmental quality. This is an outgrowth of my own proposal first introduced three Congresses ago for a select Senate committee on technology and the human environment. These proposals are based on recognition of the fact that environmental protection concerns cannot be isolated from other concerns. Membership on the proposed committee, for example, would be drawn from the several legislative committees whose activities affect the environment, committees such as Public Works, Agriculture, Interior, Government Operations, Banking and Currency, Labor, Congress, Merchant Marine, and Fisheries. Such a committee should develop a body of knowledge which would guide our legislative committees in their activities and give more visibility to environmental concerns on a day-to-day basis.

The fact is that we must build into the consciousness of every decision maker in the federal establishment as well as state and local governments an environmental consciousness. The committee of which I speak would provide a forum and a clearinghouse for those who question, those who want change and those who have ideas for the betterment of man's place in the universe. In the executive branch a more formal reorganization is needed to insure proper status for environmental protection. I am not the first to note the way in which pollution control and abatement protection programs are scattered through several departments and agencies. The Federal Water Pollution Control Administration is housed in the Department of the Interior. The Air Pollution Control Administration is housed in the Department of Health, Education and Welfare.

Now consider some of the anomalies. The Congress has assigned the responsibilities for pesticide control to the Department of Agriculture which also promotes the use of pesticides for increased agricultural production. The Atomic Energy Commission supervises radiological protection from the uses of nuclear energy which uses the commission promoted. The Corps of Engineers is responsible for some pollution control on navigable waters which the corps dredges and into which it authorizes the dumping of the polluted spoil. Some responsibility for solid waste programs are lodged in the Bureau of Mines, which has as its primary mission the promotion of mineral resource development and use. We have also given authority to the Department of Housing and Urban Development and the Farmer's Home Administration to make grants and loans for the construction of sewerage systems.

Such proliferation of activities and overlap of responsibilities are not unique to these environmental protection programs, but their proliferation is becoming intolerable because of their adverse effects on our efforts to improve the environment. So, it seems to me, the time has come to create an independent, watchdog agency to exercise the regulatory functions associated with environmental protection. Bureaus, divisions, and administrations housed in separate departments cannot marshal the resources needed to combat the interlocking assaults on our air, water, and land resources. They have neither the status nor the manpower to deal with one of the fundamental and insidious threats to our society.

I am not talking about a new department of natural resources or a department of conservation in the traditional meaning of those words because environmental protection is not the same as conservation, although sound conservation practices obviously should enhance the environment. For example, some conservation practices and projects developed and promoted by the Soil Conservation Service, the Bureau of Reclamation, or the Corps of Engineers are not consistent with broader society needs and the quality of life. Consider, for example, what we have done to southern Florida and the Everglades with our "conservation" projects in south central Florida. There is an additional reason for not proposing a department of natural resources or a department of conservation to manage environmental protection programs. The traditional concerns of conservation activities have too closely identified with the protection of natural resources separated from the population centers. Our primary concern must be man, where he lives, and the interrelationship between the natural environment and his man-made environment.

An independent agency charged with the responsiblity for developing and implementing federal environmental quality standards, supporting basic research on problems of environmental quality, stimulating and supporting research on control techniques, and providing technical assistance to state, interstate, and local agencies would reflect a national commitment we must make if we are to avoid ecological disaster. The establishment of such an agency must be backed up by a commitment of resources to eliminate the discharge of municipal and industrial wastes in our public waterways; the drastic reduction in air pollution emissions; to prevent the distribution of materials and products which threaten man and other species, and to insure the reconstruction and development of our metropolitan areas as places to live.

The commitment of resources means money and manpower, and hard decisions on where to allocate those resources and where not to allocate them. It means making environmental protection and improvement more than a conventional political issue. It is too important to be left to the emphases of the public opinion polls or the prospects of political action, confrontation, and court suits. It is too vital to man's survival to be dressed out in new committees, councils, and agencies unsupported by a willingness to invest in that survival. In the final analysis, the administration, the Congress, state and local governments will move to improve the environment in direct proportion to the degree of public awareness of the problem, the determination of the public to be heard, and the amount of informed opinion that is brought to bear on the problem. This is particularly true of those subtle threats to man's health and well-being which do not result in immediate death and obvious damage but which slowly and persistently lower our capacity to resist disease and accidents and interfere with our ability to live up to the full potential of our capacity.

Scientists have a special responsibility to society in meeting that need. We have relied on science for generations to teach us more about our world and the universe and to increase our capacity to use the resources of our planet. Now we have found that in exploiting scientific knowledge and the secrets it has unlocked, we have been exploiting ourselves. The time has come for us to adapt our scale of values and our approaches to the uses of science to man's long-term survival. The object of basic and applied science should not be to increase man's creature comforts or to overcome the natural environment but, rather, to free man from unnecessary hazards and to enable him to live in harmony with his environment. Can we implement such a concept as science and the future of man? I think the prospects are excellent. The goal of a healthy environment is an idea whose time has come.

As we look to the future, we so-called policy makers are confronted by two realities. First, it is clearer than ever before that man's survivability depends upon what he himself does to and about his environment; that the continuation of his current behavior patterns means a daily reduction in his prospects for a healthy life on this planet and that the deterioration may already be irreversible in some vital respects. The second reality is that the threshold of public patience with our failure to come to grips effectively with this problem is lower than ever before and the level of public demand that we do what needs to be done is rising rapidly. To put it bluntly, the crisis is here, the people are ready. What will the leaders do?

When I say that the people are ready, I mean that they are aware of the danger and they are receptive to a call to action. Many or most of them may be inclined to believe that someone else's behavior patterns are at fault and that the problem can be licked if someone else makes the appropriate sacrifices. But I think that most of them can be persuaded to accept restraints upon their own activities and costs which they must share. And I am led to this belief because an entire new generation, disturbed by what we are doing to their environment, is demanding that steps be taken now to protect and enhance the environment, to protect and improve man's health, to restore the balance in man's relationship to other species. They are pushing me and they will be pushing you.

Let me quote to you a portion of some remarks made recently at the Thirteenth National Conference of the United States National Commission for UNESCO. The speaker was Penfield Jensen, a graduate student at San Francisco State College. "We don't want merely to survive, we want to live! There is only one place in which to live and that is on this planet and we must live here together." I welcome Mr. Jensen and all others of like mind in his generation and in mine to our continuing struggle. It is a struggle we must win if science is to be worth advancing and man is to have a future.

My cautious optimism about the future is strengthened by the fact that we appear to have a coincidence of this problem of crisis proportions with the public receptiveness to action which is rarely found in connection with serious problems. For most of the twenty to twenty-five years during which I have been involved in legislative battles to build public policy in this field, the battles have been fought in a climate of public apathy which imposed the necessity for compromise. In writing policy, in setting goals, to use an old New England expression, "We've been moving against the wind." Now public concern is such that the wind is at our backs, and I see prospects that we can eliminate the compromises written in the current law and current policy. I can see prospects that the strength of public opinion is such as to expose, perhaps even hopefully to create, an environmental conscience not only in public decision makers but in private decision makers. And let's not forget that most of the decisions having an unfavorable impact upon the environment are in the private sector on the order of ten or twenty or thirty to one.

One of the real questions we face is whether or not it is possible to influence private decisions in such a way that they are geared to a sensitive appreciation of their environmental impact. Will American public opinion of today, for example, forestall the development of today's version of the internal combustion engine, which became the power plant for today's automobiles as the result of a private decision decades ago but which has unleashed upon our society these terrible problems of air pollution, congested cities, deteriorated urban environments? Is our system such that we can forestall or at least reshape such private decisions so that we will not see in the future irreversible results flowing from decisions over which the public interest has little or no control? This means, of course, the creation of new institutions, new forms of public policy making, new ways to arrive at private decisions, and above all, a far wider ranging and higher quality form of participation by our citizens in tomorrow's decisions.

Chairman:
In the pages of Science each week, Philip Abelson, editor since 1962, speaks to more scientists than perhaps any other scientific editor in the English-speaking world. He received his B.S. in chemistry, and his Ph.D. in nuclear physics, the latter from the University of California at Berkeley in 1939. He has had a distinguished career as a scientist and administrator of scientific research since 1939 at the Carnegie Institution in Washington. He has served as director of the Geophysical Laboratory since 1953 and has received several medals and awards for his distinguished contributions to science. As co-discoverer (with E. M. McMillan) of neptunium, 1940, and co-initiator of the study of transuranium elements, he was instrumental in obtaining nuclear fission. He is eminently well qualified to discuss here, and over the years has continued to discuss in the pages of Science on a regular basis the problems of science and the future of man.

Remarks
Philip H. Abelson

First, I would like to say a few words about Senator Muskie and his activities. I first became aware of them back in 1964-65 when I was reading some committee prints on hearings having to do with water pollution. I found that Senator Muskie had held a set of hearings across the country dealing with various problems not only in Maine but in Missouri, California, and in other states, and that the contents of those committee prints were on a high level of scientific accuracy and content. He had drawn upon the best talent in the country to give him advice in water pollution matters. He has been a great source of strength in this field, and I am delighted to hear that he is moving toward the broader field of environment.

I was pleased to hear him remark today that it wasn't a case of who owns the production but how production is managed. You know, I am a little bit conservative in these matters, having lived in Washington for about thirty years, and I am not so sure that the government can do everything for the people. I think there has to be a little bit of private initiative in this matter. My feeling on this was reinforced by something that has come out of Russia recently. In Russia, the government owns the means of production and supposedly this is all handled for the people. Yet, they have done such a thorough job of polluting their rivers that sturgeon are being killed. Thus, Russians no longer can have their caviar and, in fact, have to manufacture synthetic caviar because they have done such a thorough job of water pollution. So it is not a question of who owns production but how production is managed.

We shouldn't underestimate the willingness of private concerns to take action once their noses are pointed in the right direction. It so happens that I lived as a boy in the town of Garfield, Utah, where there was a great smelter which gave out sulfur dioxide in quantities that would make Senator Muskie pale because it was nothing like two-tenths of a part per million. There was so much sulfur dioxide in that town that within ten miles of the smelter not a blade of grass grew. I spent a year in that environment. All the trees were dead; everything was dead. So, we have had great pollution problems for some time, but an encouraging thing is that I noticed recently that Anaconda Copper Company got the message. They now have developed a wet oxidation process with the sulfur fumes completely retained and managed in the factory. The net result will be a saving of the sulfur and its use as sulfuric acid. Once it becomes clear to industry that they must take action, they will do it. Another example that I saw recently was a statement by Henry Ford which indicated that he realized that unless the motor companies really took pollution seriously they simply might be run out of business. He seems to believe that, and I think that we are going to get more cooperation from the motor industry than we have known heretofore.

I found some of the statements of Senator Muskie, but especially those of Father Drinan, to be a bit flattering and a little alarming. It seemed to me that we scientists were given an omnipotence and power that I have never sensed our possessing. Particularly Father Drinan felt we had all this political whammy so that we could take on the military and industrial complex; so that we could take on just about all the problems of society including those of the underdeveloped countries. You know, there aren't that many of us in the first place. The American Association for the Advancement of Science (AAAS), which is the one organization in this country that is concerned with all facets of science, has a membership of about 140,000. Our journal, *Science,* goes to some 155,000 subscribers. That is one part in a thousand so you can see that it takes a little bit of leverage to make that one in a thousand have all that political effect.

On the other hand, we have not been quiet. As a matter of fact, some of the first concerns that were voiced about the environment were voiced at AAAS meetings. Some of the first work and one of the first publications that got public attention with respect to air pollution was a report of the Conservation Commission of the AAAS. That was the first major, recent report in that field, and we have, in many areas, been taking action and writing reports. In fact, I believe a political scientist, a sort of genealogist of ideas, could find that most of the ideas the young people of today are espousing have been very strongly advocated in AAAS meetings for the past five or ten years. For instance, I heard over the air a statement of a young man who was decrying the manned space effort. It just so happens, about six years ago I wrote an editorial in *Science* on the manned space efforts. At that time my editorial was pretty unpopular and I was nearly fired as editor for doing this.

The other point I would make is that naturally all scientists are not alike. We do not all speak the same dialect. It's just the same with politicians. Senator Muskie is very conscientious and is very much concerned about the environment. But I am sure that we could find in the United States Senate plenty of opposition to him. Therefore, he must maneuver carefully, and you can appreciate that there are plenty of people who have differing views.

In Father Drinan's remarks he also spoke a good deal about the Third World and its expectations. It is certainly true that the people of the Third World do have great desires and great expectations. In view of that, I can't say that our policy has been entirely satisfactory, and I mean the policy of the United States government as a whole. Thus, in our AID programs what have we done? We've made a great fuss and to-do about shipping to them unwanted surplus agricultural commodities. We are regarded as a great people because we are doing that, but there has been, until recently, very little in the way of transfer of knowledge or technology. Now, I sense a change in Washington in this regard. Dr. Hannah, who has come to Washington's AID from Michigan State, has recently introduced substantial efforts to increase the transfer of science and technology. Our National Academy of Sciences is now participating in that sort of program in which workshops are conducted in selected countries and an effort is made to identify the problems that can be effectively handled, so that the local talent can be mobilized and utilized effectively. In the end we can't feed these people forever with their population explosion. We could hand out all our wealth to them and they would still be poverty-stricken. They have got to do something for themselves. We can help them, but that help, to be effective, has got to be largely in the form of transfer of knowledge. I can't quote the Chinese proverb exactly but it's something to the effect that the man who gives another man a fish gives him only a meal, but the man who gives him a fishhook gives him a livelihood. And that's what we have to do. We've got to help them to help themselves.

Chairman:
The discussants in our program were not provided in advance
with copies of the papers they were expected to discuss, in depth.
Under these conditions we were particularly fortunate to be able
to draw upon the extraordinary talents of Erwin D. Canham. Mr.
Canham is the articulate editor of The Christian Science Monitor,
a newspaper of major international importance, and is equally familiar with the problems of the local Boston community as with national and world affairs. His editorials have displayed knowledge
and concern on environmental trends and problems, and this will
be evidenced. Mr. Canham presents a familiar face for as a television moderator he has few peers, as most of us have witnessed.

Remarks
Erwin D. Canham

Mr. Chairman, when you began to introduce Mr. Abelson, you
said that you weren't quite sure what the difference was between
the speakers and the discussants. I can tell you at once what that
difference consists of. The speakers had prepared speeches. Neither Mr. Abelson nor I has that advantage, but we have had the advantage of listening to two exceedingly interesting speeches. I am
not quite sure where to begin, but I hope I will know where to end.

I was intensely interested in everything that my friend Father
Drinan had to say. He began by telling us that he fears that some
sort of massive attack or hostility toward science will emerge
from the Third World because of the great disparities which so
palpably do exist. He was concerned that a great wave of hatred
of science would emerge from the Third World, the underdeveloped world, and he was warning us of a need to reallocate our
priorities and to come to grips with the needs and the revolutionary forces which are so visibly stirring in the world. I am not so
sure that it is science or scientists which will be the foremost
target of the resentments of the Third World. Certainly, pressures

to reallocate priorities are great and inexorable. The priorities do have to be reallocated. But I am not sure that they are really scientific priorities accurately so defined. I agree with Mr. Abelson that the gap is not so much in the growth and distribution of knowledge as in the application of knowledge. I agree that the great need in the underdeveloped world is for technicians and for practical men and women who can apply the knowledge of science.

Father Drinan referred to a kind of brain drain, although he did not use that phrase. He talked about the fact that many scientists had left the underdeveloped world, that they had been preempted and drawn into the rest of the world, and this is certainly true. Again, it is complicated. I remember talking here in Boston a year or two ago with the ambassador of India and discussing with him the phenomenon by which so many students from his country get jobs and, somehow or other, prolong their stay in the United States as students as long as they possibly can, and even try to make their careers here. The ambassador said to me with complete candor and blandness, "That's all right, that's a good thing. There aren't jobs for them in India so they should stay here." Well, this feeling rather appalled me, and I think it should appall everyone. On the other hand, it's a practical situation, and the need, it seems to me, surely is to put the focus on the development of the capacity to utilize knowledge. It doesn't seem to me to be so important to identify where knowledge is generated or in what research laboratory some great new truth emerges. The problem is to apply that knowledge and to train those individuals in the underdeveloped world indigenous to that society and nation who are able to take knowledge and put it into application.

I had a luncheon visit in Boston not long ago with the director of the Peace Corps who, some of you may have observed, had a rather hard time at Harvard University, where he had been attempting to do some missionary work. I told him I thought he had gone to the wrong university, for had he gone to an institution like Northeastern University, where young men and women are putting part of their time into the application of what they are learning, or hopefully trying to, and the rest of their time in study, he would have found better prospects for the Peace Corps. He would have found individuals who have a link with the application of knowledge.

There is some very important basic research now taking place in the underdeveloped world. If one were to pick out the place where the most useful work has been done for the solution of the real problem that Father Drinan so vividly set before us, I would suppose that it is the Rice Institute at Los Baños in the Philippines. Tremendous work has been done there, as you undoubtedly all know, in development of strains of rice and other technical achievements that will make it possible to feed the hungry world much more adequately than ever before. This is already beginning to take effect.

Both ends of this process are important: the research, it seems to me, can take place almost anywhere where people are capable of doing research. The application of research has to be carried out and fulfilled by individuals who know how to apply it and who are part of the society and the culture where the application is taking place. In the case of applying the results of the knowledge developed at the Rice Institute, I suppose one of the big problems was not the actual development of a magnificent new strain of rice but how to persuade the peasants, after they had grown as much of the new strain of rice as they ever produced before, that they should continue work for the rest of the year and produce another rice crop. The tendency was to stop work after they had accomplished as much as they had ever done before in one year. There was a kind of psychological and sociological problem—as well as a scientific one—which it was necessary to solve in order to close the gap.

There is a great deal that can be done. We should reorder our

priorities, put emphasis very much more on application, and
not be apologetic as regards the location in which the research
of the pure scientist takes place. To pour financial aid into the underdeveloped world, as Mr. Abelson has hinted, is just not enough.
There is a rather cruel aphorism, with which I agree only in part,
which says that foreign aid is taking money from poor people in
rich countries and giving it to rich people in poor countries. This
is grim but there is some truth in it. It is not good enough to have
that kind of process continue. It illustrates the limitations of mere
financial aid and the importance of making aid a sort of cooperative applied knowledge.

There is so much more that one would like to say about the ex-
tremely provocative and challenging ideas which Father Drinan
set before us. He raised the question whether capitalism could
build a just world. I would answer, of course it cannot, but a just
world can be built by the cooperative efforts of the various systems.
Capitalism, as we understand it, is a system that has developed
within our own environment over a considerable period; it is evolv-
ing, changing, and must change still more. It must become much
more socially responsible and attuned to a new set of priorities
Some of this change is taking place, as Mr. Abelson says, but it
must take place at an increasing tempo, and it must cooperate
with whatever those systems are that can be operated by the people
who live in the Third World. I do not know what those systems
are. I am sure they will be mixed systems with some element, I
hope, of the thrust and vibrant dynamism of a free enterprise or
profit-motive system. I think incentive and reward must be embodied in the systems that people can operate, but I am quite sure
that nothing closely resembling the American system could be
transplanted to Peru or to the Congo or anywhere else. It will be a
system compatible with the capacities of peoples there, and it
will be a mixed system of some kind.

The United States should not be in the business of either shor-
ing up or tearing down systems in other parts of the world. We

have done too much interfering in the decision-making process of peoples in the Third World. This does not mean that we are going to withdraw and isolate ourselves but that we should be far more discreet and careful in our relationship to make sure that we are not bringing an inordinate influence upon the decisions of peoples in this vital part of the world. American capitalism, American business, can be of great value in building a just world.

Senator Muskie pointed out how many of the decisions in connection with protection of the environment have to be decisions made in the private sector of the economy. Mr. Abelson pointed out how some of this was already taking place. This is true. But there will be an obvious interaction and deeper interaction than ever before between the public and the private sectors. It may even be difficult to draw a line sharply between the public and private sectors. The relationship is not a simple one.

Let us consider now those phases of the problem that Senator Muskie has set before us. The question of rescuing and cleaning up the environment brings into focus the role of government on the one hand—national, state and local—and the role of private business on the other. I am sure there will have to be much closer interaction with many controls and disciplines that have not existed before. Above all, we have to learn how we are going to pay for it. How much of the task of cleaning up the environment is going to be paid for by the consumer when he pays his electric light bill or when he buys his newspaper? Is the newspaper reader going to pay for cleaning up the Androscoggin River when he pays for the newsprint fabricated in the pulp mill which discharges waste into that river, Senator Muskie? Or is it going to be paid for by the consumer when he pays his taxes to the local or state or the national government? It has to be paid for, and, of course, we are going to be the people who will pay for it. There surely is not going to be any set rule. To some extent, the bill will have to be paid by various elements of public finance and to some extent it will have to be paid out of our other pocket, namely, the consumers' pocket.

One more or less precise illustration of the problem has been brought to my attention recently here in Massachusetts. We have in this state—and goodness knows there are in other parts of New England—quite a number of small, rather long-standing little industrial enterprises that are in trouble and have been in trouble for quite a long time—hanging on to the edges of economic viability. Many of these enterprises are polluters, many discharge wastes into the rivers and into the air. If they are to clean up the mess which they are making, somebody has got to pay for it. If they pay for it, many of them will probably have to go out of business or leave this part of the country to go to another with alluring tax advantages, or possibly some climatic advantages and the like. For so long much of New England industry has been allured in this fashion. It is not an easy matter to decide where the responsibility lies for paying these bills and how some kind of equity and justice can ultimately be achieved. But we have to face these problems and I am sure Senator Muskie is quite right when he says that the wind is at our backs and that the time of an important idea has come.

I emphasize again the element of finance, the element of public finance. Here in Boston and in Massachusetts we are well aware of how much it is going to cost to clean up some of the waste problems, and we aren't quite clear about how to finance it. The city is carrying a heavy financial burden; the state is carrying a heavy financial burden. What form should that incomparable tool of federal aid, the federal income tax, take in terms of grants or assistance to carry this out? What should be the interaction between public and private enterprise? These are, it seems to me, among the most complicated problems that have confronted government. I do not think they are beyond the capacity of our political and sociological skills to master. There is great vital force behind the idea of environmental protection, but there is still a lot to be learned. The scientists have still a great deal to give us in terms of new technologies of waste disposal, and I am sure that there will be additional breakthroughs which will possibly help to pay the costs of cleaning up the environment.

On the whole, we are in the position of people who have had a magnificent Christmas dinner but the dishes have got to be washed! Whether the dishes should be washed by the cooks who dirtied the skillets, pots, and pans or whether the dishes should be washed by the happy loafers who ate the Christmas dinner, or in what combination these inescapable tasks which have now been reduced to the level of survival should be shared, we must now decide. I think it is fundamentally a kind of political problem—not a conventional political problem, and Senator Muskie said it wasn't—but a problem that does call for the decision-making process of people who live not only in a national society, such as our own, but who live in a world closely involved in the implications of everything that is done.

The problem of environmental quality is beginning to be important in so many other parts of the world besides our own. This is the challenge of the seventies and I don't think we are going to run away from it. I don't think we will solve it to anybody's satisfaction in a few years, but we can make some progress since our awareness is at a high level. The problem is to utilize the continuing capacity of science to cope with problems, to come up with answers and then to apply them. We must work, as I have said and will conclude by saying, to put much more emphasis on the application of knowledge rather than merely the search for knowledge. In the meantime and through it all, we must realize that man is indeed the great issue involved here; that man's relationship to his environment and the survival of man's divine birthright is what is at stake. I do not think the human race is going to commit suicide whether by nuclear devices or by poisoning the air and the water. But we have to work very hard to escape from the dilemmas to which, to some extent, knowledge has brought us. We must have the awareness that, at times, one can find wisdom in the midst of knowledge and out of the heart and the inspiration of the race we can apply that wisdom for the good of all.

Discussion Session

W. S. Joyce, S.J.: I am sure that all will agree with me that the provocative, inspirational, and informative remarks of our speakers have been perfectly balanced and rounded out by the insights and the brilliance of the comments of our discussants. We come now to what might well prove the most exciting part of our program, the question and discussion period. Before opening the program to the general audience, however, I do think it is only fair to see whether any of the panel members have any brickbats they should like to cast at each other. Do the speakers wish to exchange any comments or hostilities?

Dean Drinan: I am grateful to Mr. Abelson, for he has proved my point. I would agree that the scientists have not been as sleepy as I may have suggested, since the AAAS started its work on the environment and pollution some four or five years ago. This proves my point precisely, that if scientists get involved in the moral consequences of their own work, they can cause the revolution of which Senator Muskie spoke so eloquently today.

Question (Unidentified speaker): Taking a line from a recent movie, it seems to me what we've got here is a problem of communications. I want to make a comment and pose a question to the entire panel. There is a law not passed by Congress but articulated by Clausius called the second law of thermodynamics. What this implies for us is that there is a balance in nature to which we must be sensitive. If we are not, there really isn't much hope. I am a little bothered by the fact that so many of the panelists have so much hope in science and what it can do. I think you overestimate the capabilities of science. In a way it is unfortunate that yesterday's meeting is not combined with this one because what we have come to recognize is that we cannot go on indefinitely solving problems. The solution of one problem seems to create another.

My question is how do you address yourself to the fact that man must see himself as part of the total balance of nature. He must not proliferate himself quite so much and must see where he fits in and what he can do to reestablish the balance of nature that we now seem to be destroying. We can refer to the earth as a sort of space ship and therefore look upon our environment as something which we must constantly reconstitute. Aren't we ignoring a very important question by not addressing ourselves to this idea of overpopulation and its effects?

Senator Muskie: It is difficult to answer the question because it suggests that there is something bad about solving problems and therefore this question poses a problem. Of course it is important that we understand that there are balances in nature. I don't think there is a single one; there are many. We must understand as we speak of ecology, as the questioner suggested, that man is a part of this ecology and not something apart. A great many people who speak about ecology and the necessity for taking into account man's impact upon it seem to assume that man is not a part of it. Man is a part of it. Another thing that we ought to bear in mind is that ecology is a changing thing, not static; that man is not the only species which is responsible for change, which achieves change, or which changes balances. These changes are going on all the time. It is a dynamic balance of which we speak. Now if we believe in the Creation, and the original purpose of it all, man is the central purpose. If this is so, then man must be concerned with his relationship to the rest of nature but also of its relationship to him.

In generating man's activities we often fail to take into account the consequences of what we do, not only for other species and for the environment but also for ourselves. One of the things I try to teach my youngsters—I have five who range in age from eight to twenty—is that, as they form judgments and make decisions, they must learn to project the likely consequences of what they do. These consequences may have a greater bearing on what their decision ought to be than their immediate objective—whether it is marriage, or choosing a college, or a career. They must try to project the long-term consequences. One of the great failings of mankind is, if not his inability, at least his failure to measure fully the consequences of what he does. This is difficult enough in one's private life, where one makes decisions affecting one's own activities. But when we get down to the problem of the decisions of masses of people in society, in the country, in the city and the community, it is more difficult to focus the attention of all upon the consequences of every decision, private and governmental, which is made that affects the environment of the community and society in the long run. How many Americans had anything to do with Henry Ford's decision to build the Model T? Very few, yet the consequences of what he did were far reaching.

So our challenge here, only a part of the challenge of the individual developing the ability to project consequences and understanding to react responsibly to them, taking into account his own decision, is a question of how our society organizes itself. In order that consequences stemming not only from general decisions but from individual decisions may be comprehended, appreciated, and taken into account, policies must be framed within our society. They are not all on one level. They are not all in the governmental sector. They are not all in the private sector. They are very difficult, the most difficult challenge that has ever faced man. At the turn of the century, my little town of Waterville, Maine, which is just twenty thousand people today, celebrated one of these municipal anniversaries of which so many communities have so many. The principal speaker was the president of Brown University and one very eloquent line with which he closed his speech was this one, "Americans have succeeded nobly in founding states but they have not yet learned to govern cities." We have not yet done so, and we are much more urbanized in America today because we haven't yet found the institutional ways in which to measure, identify, and take into account and respond to the consequences of the many activities that are undertaken by people in the free society.

Doctor Abelson: I would just like to comment briefly on this matter of *problems.* I believe that one of the reasons why some of the members of this panel are optimistic is because they have lived through a great many crises. They have observed that humanity is crisis-prone. If there isn't a great set of problems at hand, humanity goes about the business of creating problems. Beyond that we have also found that problems, in general, are solvable. Having participated in one way or another in the solution of a great many problems, people become relaxed about the seriousness of problems. Some of these problems have been quite difficult, but having cut their teeth on some tough ones they are ready to set them up in the next alley.

Mr. Canham: We have been laying a lot of emphasis on how much we know. It might be just as well to turn the coin around and emphasize how much we don't know. I detect among many scientists today a good deal of that healthy attitude. In fact, I think that is the primary difference between scientists of the nineteenth and the twentieth century. In the nineteenth century, scientists for the most part thought that they had learned nearly everything there was to be learned, that they had gotten to the top of knowledge. But I don't think this bland attitude of egotism really exists today. It may have existed on into the decade of the thirties—and some people then were pretty positive as some are today—but let us remind ourselves that we could be wrong about a lot of things.

Remember the famous article in *Harper's Magazine* written by then-President James Bryant Conant of Harvard University who said that population increase had stopped; that schools were not going to expand; that there were not going to be any further educational problems; that buildings would have to be replaced but nothing new had to be done in education. At least he said words to this effect. Of course, President Conant, a wise man, discovered in a little while how wrong he had been.

Maybe we are wrong now. I can vividly remember the pundits of the thirties who said things as we reached the end of the frontier: the population isn't going to increase and perhaps may decline in the United States, and so on. I am not saying that we are going to be as wrong now as we were four decades ago, but I don't think we are right now about a lot of things. Perhaps this state of healthy awareness of the possibility that the crystal ball is murky will help us as we move forward. The question of the population problem is one we have not said much about. We are not trying to run away from it. You all know how very sensitive the population problem is, and how much has to be done to cope with this terrible race between population growth and economic growth. Nevertheless, a state of healthy uncertainty about just what the future holds and a continuing questioning of premises today may be the most healthy scientific attitude. It is a stance not unfamiliar to science. If I read the history of science at all accurately, there have been ups and downs on this point through the years. Maybe the time is coming when we want to be more vigorous in asking questions than we are in assuming we know all the answers.

Question (unidentified speaker): In view of the Soviet Union's forcible annexation of all its states in 1940, the attack on Finland in 1939-40, the suppression of the East Berlin workers in 1953, the inspiration of the Korean Pact in 1950, the suppression of the Hungarian national revolution in 1956, the threat to Poland in the same year, the invasion of Czechoslovakia in 1968, the threat to West Germany the same year, the threat over the years to Yugoslavia, West Germany, of course West Berlin, even Austria; in view of the present doctrine allowing the Soviet Union to interfere anywhere they care to impose their brand of socialism; in view of the present occupation character of Germany, Poland, Czechoslovakia, Hungary, and to some extent, Bul-

garia; and in view of the present threatening posture toward Israel, what makes you think, Father Drinan, or anyone else on the
panel, that unilateral as contrasted to bilateral or mutual disarmament will bring forth a better, more peaceful, environmentally
sounder and more moral world?

Dean Drinan: I believe, like Pope John, that we ought to talk
with these people even though they have done all those things.
All that we can say, and all that we should say, is that we are in
good faith and ease our armaments in this particular way—not en-
dangering our population. The whole literature of disarmament
demonstrates that the bilateral or mutual approach just doesn't
work, and that a bold and daring step is necessary to move toward disarmament. It seems to me that unilateral disarmament is
really the only option open.

No one is denying the litany of horrors you recite. But at the
same time, can we go on in this escalation of terror, this coexistence in anguish by which we are starving literally half of humanity?

Senator Muskie: I wonder if I may make just one comment. Of
course I did not write Father Drinan's speech. Nevertheless, I
was provoked by his endorsement of unilateral disarmament as
a viable political concept. I am sure he did not have in mind the
view that we did not face in the Soviet Union an agressive government pointing toward supremacy of one kind or another in the
world. Ideas along this line are less militaristic than they were a
quarter of a century ago, but in all our dealings with the Soviet
Union I think we must still negotiate at arm's length and assume
that the Soviet Union's intentions are aggressive, whether mili-
tarily or otherwise. We must be alert to our own national interests
constantly. Nevertheless, I think the only viable way to approach
the future of this planet is to assume that change takes place on
the other side of the iron curtain as well as this side, and that national interests, especially in such vital areas as national secur-

ity, can be accommodated not out of consideration for the other fellow but out of an enlightened view of one's own security interests. I don't think that either the Soviet Union or the United States considers continued escalation of nuclear arms to be in our respective national interests. Obviously this is so, or we wouldn't be meeting in Helsinki or Vienna.

In dealing with this problem of de-escalating arms, the United States to some degree must be willing to take the risk of unilateral action from time to time. Not total, immediate, unilateral disarmament—I am sure that Father Drinan did not have that in mind—but unilateral actions which will focus the attention of both countries on the possibilities of stabilizing some aspect of the arms race.

For example, in the development of MIRV, both countries are conducting tests, and we assume that we are farther ahead than the Soviet Union. We are not sure of that, but we assume we are, and the Soviet attitudes are based on the same assumption that we are ahead. In any case, our tests are going to be completed at some point in 1970, we are told that's the timetable, so that we will be in a position to deploy the weapon. It seemed to me, two or three months ago, that this was an ideal situation to take a unilateral step toward disarmament without too great risk to ourselves. So I recommended the unilateral cessation of further tests on our part. We were ahead, we could complete the tests any time we decided to do so, but our unilateral action in suspending tests, it seemed to me, would give credibility from the Soviet point of view to our expressed desire to stabilize the arms race. It was the kind of step that we were more likely to take than the Soviet Union. After all, we are the country that identifies itself with peaceful objectives in the long run.

Other opportunities of this kind may emerge from time to time. We ought to be willing to identify them, to recognize them, and to take intelligent, unilateral steps designed to stabilize the arms race. As we develop our credibility and mutual confidence in this way, what we may see emerging is not a world in which the Soviet Union patterns its policies upon our wishes or the reverse but a world in which each of us will see increasingly that our respective national interests require accommodation of objectives from time to time.

Question (unidentified speaker): Mr. Canham partially discussed a problem in which I am most interested, that of financing the solutions to our environmental problems. I would like to address this question to Senator Muskie but would welcome comments from the other panelists too.

Those of us who have been interested in environmental problems have certainly been frustrated by the lack of seriousness with which some business executives approach environmental problems. Do you believe that the primary responsibility for the financing of solutions to environmental problems should be from the public sector? Or would you be willing to have means, such as corporate taxes levied in a punitive fashion, used to help solve and finance our environmental problems?

Senator Muskie: I am more interested in cleaning up pollution than I am in punishing somebody. Under the present scheme of things, the financing of waste treatment facilities is divided, of course, between the private sector and the public sector. The public sector, at the present time, is concerned wholly with financing municipal waste treatment plants that deal with municipal wastes of one kind or another. Under our water quality laws, for example, the standards that are imposed at the state and local level are imposed not only upon communities but on industrial plants as well, and they are required to meet those standards at their own expense. At present, there are some states and a few municipalities that exempt pollution control facilities from local taxes or from state use taxes, on the theory that these are not productive and we ought not tax these facilities that are built. But the facilities must be built at private expense. In the tax bill just passed in Congress, there is included a provision in the form of accelerated depreciation as an incentive for private industry to build pollution control facilities. It is not a heavy or massive aid, but it is an incentive that, it is hoped, when combined with the whip of enforcing standards would result in gradual improvement or compliance with the requirements of public policy. The hitch of it all is the willingness of state and local governments to impose tough standards. If you develop tough standards, it could mean, for example, the closing of the marginal kinds of small plants to which Mr. Canham referred, or the rejection of new industrial development. Local people regard this as incompatible with their own environmental protection. If you have tough standards, then I assume you also have the public willingness to pose punitive taxes. The question is which is more likely to work.

I think that tough standards which could generate various forms of financing are preferred. The idea of using taxes to force industries to build facilities comes, I think, out of the Ruhr valley in Germany. I went to the Ruhr to study what is done. They created two quasi-public institutions, one dealing with water supply and the other with water quality. Everyone who uses water or discharges water into public waterways must be a member of these quasi-public corporations. These corporations are charged by the states with the responsibility of maximizing the water supply by the construction of appropriate engineering works, the cost of which must be born by the members. In the same way, waste treatment plants are constructed by that corporation, and construction must be paid for by members in proportion to the load which they contribute. Out of that formula has come the concept of taxes as a punitive way to force the cleanup of waters, and I suppose it is. But what we are talking about really is a willingness on the part of the public, legislative bodies, people whose jobs and economic future may be involved, to set high standards of performance.

Let me tell you how easy it is to espouse these objectives and how difficult it is to implement them. When we passed the Water Quality Act in 1965, Secretary Udall had the responsibility of laying the guidelines for the setting of standards by the states. What we had in mind was a turning-of-the-screws operation; that is, the first step is taken bearing in mind that we are going to turn screws as the economics and the technology and the situation permit. Mr. Udall came to see me to ask my advice, and I said that one standard that surely ought to be in your guidelines, in the very first paragraph, is that no waterways shall be permitted to deteriorate below their present level of quality. Simple idea? Well, that one standard generated the most controversy of any. Where? In the less-viable economic areas of the country; the areas that felt they needed economic growth to provide jobs, to create opportunities, to build schools, and to provide public services which lagged behind those in other more viable areas; areas that had these marginal little industries that move into places where wage levels and public standards are low; areas, in other words, where human resources and natural resources bear a disproportionately high share of the costs of operations.

How do you convert public attitudes in these areas? It's relatively easy in the suburbs of Washington, which are among the wealthiest and the highest-cost-of-living areas in the country, to build up a public motivation for tough quality standards, but it's much more difficult in these other areas. That is the nature of the problem we have, and this is why public apathy has resulted in the despoiling of the pristine pure waters of our country.

It's beginning to spread like a contagious disease. Even in Maine we are beginning to get pullution of our waters and our air because of the urge of people in economically less viable areas of the country, relatively, for economic growth and opportunity. This is the lesson we have to learn. It's to that kind of urge that the less responsible and dirtier, in the environmental sense, industries move.

You've learned by now that a question prompts a senatorial speech.

Comment (unidentified speaker): What you just said, if slightly generalized, will also apply directly to the problem of population explosion. What do you do in India to make them less motivated to have large families? Could you start taxing them for more children? Who could stay in power very long with the population that doesn't understand that this is what they must have? They would have to have a sophisticated understanding of what is in store in the future because of the constant rate of population expansion. They don't realize that they are at the edge of a cliff. How can you communicate to them effectively so that you would be allowed to do what is absolutely necessary? Taking your remarks on pollution over to this problem, it almost seems as if there is no way out, as if the system is doomed. I do not accept that, of course, and would like you to say something.

Senator Muskie: Well, education, enlightenment, and communication, of course, provide the answers.

I first went to the Maine legislature in 1947, and since that time I have seen, for example, a great change, almost a miraculous change, in the public's attitude about pollution. Just this last year, I think Erwin Canham may remember this, on the coast of Maine in the little town of Trenton—up near Bar Harbor but not in that wealthy complex, it's one of the less-developed, less-viable rural areas of Maine—the people rejected in referendum an aluminum plant that promised jobs, jobs badly needed I might say, because of the environmental considerations. So you see, I think there is a rising tide of public concern across the country to which I referred in my prepared remarks. I feel that the wind is at our backs now on this subject. You get the same kind of enlightenment with respect to population. But it's going to take education. You can't put the government in every bedroom of the planet. And taxing people for creating children is like taxing people for failing to vote. What you want is positive action, not repression. You can't control the regenerative forces of human nature with taxes. You have to do it by working at the other end of the human being, his mind and his understanding of life. In this day of unparalleled development of technology and communications, it seems to me that man is in a better position than ever before to perform that job. But I suppose I ought to leave that not to the politicians but to the teachers, preachers, and those who can really reach inside people to create true enlightenment.

Comment (unidentified speaker): I am a teacher of environmental science to thirteen and fourteen year olds and I hope they can become leaders in the twenty-first century. I have a comment since the question I had was just answered by Senator Muskie. I would like to suggest another valuable contribution, perhaps the most valuable contribution, that scientists can make. In the discussion today, it seems that scientists have been viewed more or less as the servants of society who can come up with the technological solutions to some of the crisis situations that have arisen. That contribution is this: By virtue of the intellectual discipline that scientists practice, they are able to break out of the man-centered universe and world unisonance which so many of us hold and seek patterns in broad prospective. A consequence of this holistic output is not just a grave concern with the environmental deterioration but a very great fear. I would like to say, and this has come of yesterday's session, that to those on the outside this often appears as hysteria. I hope, however, that people will not view it as such in the long run.

Question (Paula Lyons, Boston College): I would like to ask Senator Muskie a question on the grand old institution of home rule and how it often affects regional cooperation for solution to environmental problems, transportation, and so on. I wanted to ask you, Senator, if you can see federal legislation that would encourage, if not compel, the states to take a regional river basin approach to pollution abatement and, in general, to water resources management—an approach that has proved more efficient and economical than our current town by town, industry by industry, piecemeal approach.

Senator E. Muskie: It is our intent in the Water Quality Act and in the Air Quality Act to stimulate the development of regional approaches. This would not necessarily result in regional institutions, but it certainly requires regional cooperation. The water quality standards of a given stream obviously spill over local jurisdictional lines, and they ought not to be approved by the secretary in Washington unless the total river basin has been evaluated. This is the whole purpose.

Now it is our hope that there may develop regional institutions to deal with these problems. As a matter of fact we have gone beyond this in some of the federal legislation with which I have been associated—in the housing field, in the urban renewal field, and the transportation field. For example, we've undertaken the creative centers for the development of regional institutions in metropolitan areas, and we've undertaken to stimulate the development of regional planning as the very first essential in the development of metropolitan-wide institutions. We are faced with a dilemma here because in this day of participatory democracy, people are very sensitive to the development of new and larger institutions. How do we develop metropolitan-wide regional institutions as broad as the problems with which they must deal and yet keep those institutions close to the individual citizen who feels a part of them and as a consequence will participate and undertake to shape their decisions? This is the great challenge. This is the new institution-growing challenge that we face in the metropolitan America of tomorrow. And I think that the river basin and the air shared, which God has certainly created without reference to political boundaries, are two areas of public concern that are ideally suited to the development of new regional concepts of government. If this doesn't develop, then what we may see developing is a stronger national role, a stronger federal role, with the regional problems handled by federal agencies.

In our committees we did not like that idea. We tried to steer developments in the opposite direction because we think, from an administrative point of view, as Mr. Abelson said, that the government in Washington cannot do everything. We hope to stimulate the development of regional institutions springing out of state and local governments rather than imposed by federal bureaucracy as part of that bureaucracy.

Question (unidentified speaker): I think it is unfortunate that we are picking on Senator Muskie, but I would like to ask him whether this is really a scientific problem totally or whether it is very much a political problem. Many of us in science feel that a great many of the answers have been provided already and the technology is available at least to begin on the preservation and restoration of our environment. The United States, with less than one-tenth of the world's population, consumes three-tenths of its resources, and we should therefore be a leader in the field of preserving and restoring the environment. This needs, of course, a new set of economic priorities because we won't have enough money to build SSTs and ABMs and clean up the environment at the same time. And against scientific advice some of these other projects have been continued while some of the environmental ones have been discouraged. As a general citizen, I just wonder if there is going to be a change in viewpoint with respect to economic priorities of the federal government.

Senator E. Muskie: I really don't want to make all these speeches, but your questions don't leave me much choice. First of all, there is no question that there is a lot of technology available today which if it were used and applied could very quickly make a difference in the quality of our environment. On the other hand some of it produces new problems. Most municipal waste treatment plants in operation now, for example, provide primary treatment, some provide secondary treatment, and very few, if any, provide tertiary treatment, which would restore the water to drinking quality. At the same time these treatment plants discharge nutrients into the waterways which stimulate the process of eutrophication. So the great technology that cures part of the problem creates another one. We must constantly work for breakthroughs.

In another vital area technology is inadequate: with sulfur oxides. In 1966, the administration asked us to enact legislation to create national emission standards as a way of controlling air pollution. We didn't, and one of the reasons we didn't is because we don't have technology to deal with sulfur oxides. The only way we have of dealing with sulfur oxide effectively today is by fuel substitution, using low-sulfur content fuels. The supply is relatively limited, and we felt this fuel should be available in the problem areas. But if we were to have national emission standards, it would mean that we would have to use this limited supply wherever sulfur oxides were produced and generated by the manufacturing plant. We wouldn't have enough for the problem areas. So again technology is the handicap. Nevertheless, I reemphasize, if we use all the technology that is now available, we could make a visible difference, very quickly, as soon as you could make the investments and install the technology, in the quality of our air and our water.

Whether or not we are moving in the direction of changing priorities I think we have too little evidence to conclude. Father Drinan touched upon this—all of the panelists did. In this session of Congress I feel we have at least made a significant first step. For example, we cut the administration's military budget by something on the order of 6 or 7 billion dollars. We increased the recommended budget in such matters as water pollution by $600 million and education by $1 billion. Each of these differences in judgment provokes controversy with the administration, which is *fine* because this focuses public attention on what we are trying to do. There will be some who disagree with the Congress, some who disagree with the president. But I think these differences have been significant enough and massive enough to suggest that the country is moving in a different direction in the matter of priorities. One of the most significant changes of all is the change in attitude of Congressman Mahon, who is chairman of the Appropriations Committee of the House. He is from Texas, and some of his attitudes are readily associated with his state. In any case, he is leading the fight now to cut military spending. That is a significant change.

Question (John J. Maguire, Boston College): I agree that Senator Muskie has been beset by too many questions so, after a brief comment, I would like to ask Dr. Abelson a question.

It seems a bit ironic, to me at least, having sat through every session for the last day and a half of a symposium on science and the future of man, that not one mention has been made at this symposium of what I consider one of the most important documents ever produced by a scientist precisely on the question of the future of man: the Sakharov papers. These were produced by Andrei Sakharov, the Russian father of the hydrogen bomb.* He addresses himself to many of the questions that have been touched upon in part by Father Drinan with regard to our relations to the Third World, and in part by Senator Muskie with regard to the crisis of our environment. He quantifies in a little less detail some of the problems that John Platt alluded to in his fine talk yesterday. For those of you who haven't read the Sakharov papers, I recommend them strongly. They present a fantastic point of departure in any discussion that scientists, economists, and politicians might undertake to deal with some of the really testy world problems and some of the unthinkable thoughts that Father Drinan talked about, like world government, the diversion

---

* Andrei Sakharov, *Progress, Coexistence, and Intellectual Freedom* (New York: Norton, 1968).

of great resources into the Third World by the developed coun-
tires, and many other questions. I think these are some of the unthinkable thoughts that we have to begin to think about, as scientists.

My question is to Dr. Abelson. I am sure it has been asked of you
many times before, but I have never heard your answer. How far
do you think professional societies such as APS, AGU, and AAAS
should go, to what lengths should they become involved in political processes, in political lobbying, in political decision making?
What is your feeling exactly on whether or not the force of scientific opinion through societies would in any way be diluted by
their active involvement in political questions?

Doctor Abelson: Scientific societies have a number of responsibilities, but as you know they were originally created to serve
some special needs of the scientists for communication, that is,
to organize meetings such as this symposium and also to organize
the publication of scientific findings. Now, if the professional societies don't take care of those functions, nobody else will. On the
other hand, there are all kinds of mechanisms for handling a political problem. I believe that scientific societies should remember that there are these functions that they alone can serve
and that in engaging in political activities, which I believe they
should do to a certain extent, they should not proceed in such a
way as to fragment the societies. It is quite possible for scientists
to get so mad at each other as to destroy the organization itself.
That kind of devisiveness would not, of course, bring any useful
political activity.

On the other hand, I think that the model that you can see in
the activities of the AAAS is a pretty good model, namely: the
AAAS serves as a platform for the discussion of these problems
of society. As a result, there is a gradual mobilizing of public
opinion for action. This is accomplished not only through meetings as this symposium but, for example, in our own magazine,

Science, where many scientists, such as John Platt, publish their material and receive wide circulation and attention. John Platt's article is now in the *Congressional Record.* Many of the items that appear in our magazine subsequently appear in some version in the *New York Times* and in the national weeklies. There is an enormous amplification effect here that is too valuable to hazard destroying by intense preoccupation with a single issue. We must remain a viable organization in order to have these large amplification effects.

W. S. Joyce, S.J.: I am sure that our panel wishes to express their gratitude to the audience for your very kind attention and provocative questions: I know you wish to thank the panel for their splendid presentations and insights. It was a pleasure to be with you. The meeting is adjourned.

Name Index

Abelson, Philip H., 159, 163, 164, 166, 167, 171, 174, 184, 187
Asimov, Isaac, 124

Bacon, Francis, 103-104
Bade, Maria, 64
Becket, Thomas à, 35
Bethe, Hans, 96
Blackett, Patrick M. S., 85, 86, 88
Bohr, Niels, 96
Born, Max, 112
Burton, Richard, 81

Canham, Erwin D., 163, 174, 178, 179
Carmichael, Stokely, 39
Carovillano, Robert L., 121, 125, 131
Chrysler, Walter, 8
Comte, Auguste, 106
Conant, James Bryant, 174, 175
Copernicus, Nicolaus, 8
Cox, Harvey, 84

Darwin, Charles, 8
da Vinci, Leonardo, 111, 112
Descartes, René, 104, 105, 126
Deutsch, Karl, 85
Drinan, Robert F., S. J., 134, 161, 162, 163, 164, 165, 166, 171, 176, 177, 185, 186

Einstein, Albert, 8

Faulkner, Robert, 127
Feynman, Richard P., 11
Ford, Henry, 8, 160, 173
Freud, Sigmund, 111
Fromm, Erich, 144

Galbraith, John Kenneth, 144
Galilei, Galileo, 8, 104, 126
Glass, Bentley, 12
Gordon, Clancy, 25
Guevara, Che, 21
Guthrie, Woody, 34

Hannah, John A., 162

Hays, Harry W., 30-31
Hooke, Robert, 110
Hornig, Donald F., 50
Hugo, Victor, 118

Inglis, David, 87

Jensen, Penfield, 157
John XXIII (pope), 137
Johnson, Lyndon B., 50
Joyce, W. Seavey, S. J., 133, 171, 188

Kennedy, John F., 50, 137, 138
Keynes, John M., 85, 86
Kimball, George E., 88
Koestler, Arthur, 8

Leefert, Robert, 58, 59
Lincoln, Abraham, 79, 145
Long, Franklin A., 72, 91, 97, 113, 117, 119, 120, 124
Lyons, Paula, 182

McLean, Louis, 62
McMillan, Edwin M., 159
McNamara, Robert, 83
Maguire, John J., 57, 186
Mahon, George H., 186
Mansfield, Mike, 52
More, Thomas, 35
Morgenstern, Oskar, 86
Morse, Philip M., 88
Mumford, Lewis, 103, 120, 125, 126, 127, 128, 129, 130
Muskie, Edmund S., 147, 159, 160, 161, 167, 168, 169, 171, 172, 176, 178, 181, 182, 184, 186

Newton, Isaac, 8, 104, 126
Nobel, Alfred, 143

Orwell, George, 97
Osgood, Charles, 144

Parks, Paul, 36, 66
Pastore, John O., 101
Paul VI (pope), 138

Pauli, Wolfgang, 96
Pauling, Linus, 87
Platt, John, 77, 79, 91, 97, 99, 102,
    113, 114, 116, 117, 119, 120, 121,
    122, 123, 124, 126, 128, 186
Polanyi, Michael, 98
Ponnamperuma, Cyril, 59

Quelle, Fred, 60

Risebrough, Robert, 62
Roberts, W. O., 142
Roland, Henry Augustus, 148
Ruml, Beardsley, 86, 87

Sagen, Carl, 113
Sakharov, Andrei, 186
Schindler, George, 62
Skehan, James, S. J., 1, 59
Snow, C. P., 83, 104, 142
Spencer, Herbert, 106
Sternglass, Ernest, 60-61

Thant, U, 83, 84
Thayer, George, 144
Truman, Harry S, 138

Udall, Stewart, L., 180

von Neumann, John, 86, 88

Wald, George, 91, 100, 102, 114,
    116, 117, 118, 119, 120, 121, 122,
    123, 124, 129, 130
Ward, Barbara, 139
Weisskopf, Victor F., 96, 103, 116,
    118, 120, 121, 123, 128
Wigner, Eugene, 96
Williams, Curtis, 65
Wilson, J. Tuzo, 3, 57, 58, 59, 60
Wilson, Robert, 100, 101
Wurster, Charles F., 62

Yannacone, Victor J., 20, 24, 60, 62,
    65, 67

Subject Index

ABM (anti-ballistic missile), 76, 101, 115, 116, 135, 184

Abortion, 47, 134

Academic freedom, 16

Advancement of Learning, The, 103-104

Advertising, 119

Africa, 138

Agriculture, U.S. Department of, 23, 30, 31, 35, 51, 67, 153

AID (Agency for International Development), 140, 162

Aid for Dependent Children (AFDC), 48

Air Force, 86

Air pollution control, 22, 25, 49, 60, 67, 68, 148, 158, 160, 185

Air Pollution Control Administration, 153

Air Quality Act, 183

Al Fatah, 140

Algae, 22

AMA (American Medical Association), 46

Amazon, 64

American Association for the Advancement of Science (AAAS), 64, 79, 142, 148, 161, 187

American Trial Lawyers Association, 63

Anaconda Copper Company, 160

Androscoggin River, 167

Antarctic, 81

Antarctic Treaty, 123

Apollo Mission, 74

Applied science, 88, 90, 124

Armaments, 115, 116, 144, 176, 177

Arms race, 74, 177

Army Corps of Engineers, 23, 29, 31, 35, 62, 153, 154

Association for the Integration of the Sciences and Humanities with Life, 103

Atlantic Ocean, 149

Atomic bomb, 80, 99, 112, 143

Atomic Energy Commission (AEC), 23, 35, 51, 52, 60, 61, 63, 153

Automation, 107

Banking and Currency Committee, 153

Basic research, 73, 75, 78, 79, 88, 89, 96-102, 121, 124, 136

Basic science, 5, 96-98, 120, 121-122

Bedford-Stuyvesant, 68

Belgian Congo, 140

Big Thicket, 64

Bioscience, 62

Birth control, 47

Black Muslims, 139

Blacks, 21, 22, 39, 66, 67, 139

Boston, 40, 163, 164, 168

Boston College, 56, 64, 127, 134, 182, 186

Boston College Law School, 134

Boston College Environmental Center, 59

Boston Globe, 91

Brain drain, 164

Brazil, 64-65

Brown University, 50, 124, 173

Bulletin of the Atomic Scientists, 87, 144

Bureau of Mines, 154

California, 34, 159

California Institute of Technology (Caltech), 87

Cambodia, 116

Cambridge University, 87

Canada, 15

Cancer, 82, 89

Capitalism, 143, 166, 167

Capitalist system, 117

Carnegie Institute, 159

Center for Environmental Quality, 59

CERN (European Organization for Nuclear Research), 96

Chicago, University of, 87

Christian, 4, 108, 136

Christian Science Monitor, The, 163

City, 37, 38, 39, 40, 49

Commerce, 115, 117

Communication, 80, 82, 117, 183

Communism, 138, 143

Compartmentalization, 57

Computers, 80, 82
Congress, U. S., 25, 26, 27, 113, 114, 152, 179, 185
Congressional Record, 188
Conservation Commission of AAAS, 161
Conservation movement, 23, 24
Constitution, U.S., 127
Constraining science, 7-8, 15
Contraceptives, oral, 85, 86
Cornell University, 72
Costs, 155, 167-168, 178-180
Credit cards, 85
Czechoslovakia, 41, 175

Data handling, 80
DDT, 22, 29, 30, 36, 49, 61, 62, 67, 81
Declaration of Independence, 145
Defense contracts, 115
Defense, Department of (DOD), 51, 52, 149
Democratic party, 114
Denver, 149
Delaware River Basin, 149
Department of Defense (DOD), 51, 52, 149
Department of Health, Education, and Welfare (HEW), 153
Department of Housing and Urban Development (HUD), 56, 154
Department of the Interior, 56
Deuteronomy, 95
Disarmament, 14, 16, 74, 144, 175-178
Discourse on Method, 105
Discovery in science, 11
Diseases, 80, 82
DNA, 53, 56, 73, 136
Drugs, 129

East Harlem, 68
Ecological systems, 22
Ecology, 22, 100, 151, 172
Ecosystem, 23, 24, 30, 65
Education, 41-44, 181-2
Energy, 80, 82
Engineers, Army Corps of, 23, 29, 31, 35, 62, 153

Environment, 16, 23, 25, 28, 29, 30, 32, 33, 34, 35, 37, 60, 94, 97, 99, 100, 120, 135, 147-158, 167-169, 178-180, 184
Environmental control management, protection, or quality, 22, 23, 25, 26, 27, 28, 29, 30, 33, 36, 77, 81, 151, 152, 153, 154, 155, 168, 169, 184, 185
Environmental law, 63
Environmental Quality Council, 152
Erindale College (University of Toronto), 3, 59
Establishment, 8, 96, 138, 141
Everglades, 31, 32
Evolution, 81, 91-95, 100, 127, 117, 155

Faith, as trait of scientists, 4-5
Farmer's Home Administration (FHA), 154
Federal Power Commission (FPC), 28
Federal Radiation Council, 67
Florissant, Defenders of, 26
Florissant fossils, 25-27, 31, 32
Food and Drug Administration (FDA), 67
Ford Foundation, 135
Foreign aid and policy, 138, 140, 143, 162, 165, 166
Future of man, 148, 156, 157, 186

Game theory, 86, 88
General Motors, 33, 35
Generation gap, 130
Germ warfare, 148
Germany, 99, 179
Grades, 58, 59
Grand Canyon, 29, 32
Great discoveries, 10, 11
Growth rates, 3

Harlem, 68
Harper's Magazine, 174
Harvard University, 50, 91, 103, 164, 174
Head Start, 42

Health, 44-46, 56
Health care, 44, 45, 46, 51, 53, 54, 56, 66, 67, 82, 89, 99, 149
Henry Luce Professor of Science and Society, 72
Hippies, 113, 114
Historical perspective, 84, 97-98
Hot Line, 123
Housing, 22
Housing and Urban Development, Department of (HUD), 56, 154
How to Control the Military, 144
Humanities, 104, 105, 111
Hunger, 51, 99, 165

Income tax, 86, 87
India, 164
Indian, American, 22
Industrial pollution, 151, 155
Industry, 160
Imperialism, 115
Institute for Advanced Study, Princeton University, 87
Intercontinental missile, 139
Interdisciplinary fields or programs, 15, 17, 57-60, 76, 101, 104, 119, 149
Interior, Department of the, 56
Internal combustion engine, 158

Japan, 13, 129
Jetport, 63

Keynesian, 87, 118

Laos, 116
Latin American, 138
Law, 20, 21, 22, 23
"Law and the Urban Crisis," 21
Leaders, failure to train, as trait of scientists, 8
Leonardo da Vinci Medal, 110
Lead poisoning, 66, 67
Litigation, 20, 25, 28, 32, 33, 61
Living Theater, 109
LSD, 129

Madison, Wisconsin, 30
Maine, 147, 159, 180, 181

Manhattan Project, 99
Man in space program, 79, 89, 161
Mansfield amendment, 51-52, 54
Marijuana, 129
Massachusetts Institute of Technology, 96
Military, 4, 6, 8, 161
Military industrial complex, 161
Milky Way, 93
MIRV (Multiple independently targeted reentry vehicle), 101, 115, 135, 144, 177
Missouri River, 29
Modern Cities Program, 36
Moon landing, 103
Morality, 5, 46, 47, 48, 120, 134, 141, 143
Mount Palomar, 93
Myth of the Machine, The, 111

NASA (National Aeronautics and Space Administration), 51
National Academy of Sciences, 162
National Audubon Society, 31
National Environmental Policy Act, 152
National Humanities Foundation, 135
National Science Foundation (NSF), 51, 52, 55, 135
Natural resources, 12-13, 15, 23, 25, 29, 31-33, 35, 69, 80
Nature of science, 9, 72-73, 103-104, 110
Nazi, 86, 99
New York, 24, 29, 34, 40, 68, 148
New York Daily News, 68
New Deal, 86, 114, 118
NIH (National Institutes of Health, 51, 52, 53, 88-89
Nile, 81
1984, 97, 128
Nobel Prize, 74, 91, 143
Northeastern University, 164
North Pole, 81
North Viet Nam, 116
Nuclear Non-Proliferation Treaty, 86, 123
Nuclear Test Ban Treaty, 86, 87,

Nuclear Test Ban Treaty (continued) 123, 141
Nuclear testing, 60, 61

Ocean, 94
Office of Environmental Quality, 152
Oil pollution, 149, 151
Oligocene period, 25
Oral contraceptive, 85, 86
Overkill, 82
Overpopulation, 22, 50, 72, 80, 82, 89, 135, 172, 174, 175

Pacem in Terris, 137
Patriotism, 5
Peace, 77, 78
Peace Corps, 164-5
Pentagon, 125, 144
Pentagon of Power, The, 105, 126
Pesticides, 22, 23, 30-31, 61, 62, 63, 64, 153
Pesticides Registration Division, 30
Photosynthesis, 94
Pittsburgh, University of, 60
Pogo, 78
Point Four, 138
Poland, 175
Polio, 89
Polio Foundation, 89
Political action, 115, 116
Political institutions, 113
Political structures, 39, 40
Pollutants, 28
Pollution, 12, 23, 25, 36, 50, 54, 56, 58, 60, 67, 68, 74, 77, 78, 81, 87, 94, 99, 107, 114, 135, 148-158, 160, 168, 183, 185
cost of controlling, 168-169, 178-180
Population, 77, 78, 80, 82, 90, 151, 174-175, 181-182
Poverty, 22, 51, 74, 77, 78, 120, 141
Power, 125, 138, 140, 161
Princeton University, 3, 50, 87
Priorities, 164, 166, 184-185
Priorities of science, 7, 14-16, 135
Private initiative, 158, 160, 167, 178

Professional societies, 187-188
Profit, 114, 117, 125-126
Prophets, 130, 131
Program on Science, Technology and Society, Cornell University, 72, 78
Project Rulison, 61
Public attitude and concern, 76, 97, 156, 157, 167, 181
Pugwash Conferences, 87, 123

Radiation damage, 99-100
Rand Corporation, 86
Regional government or approaches, 182-183
Religion, 4, 6, 8, 9, 95, 134-135
Republican party, 114
Research, 5, 10, 13, 17, 51-56, 88-90, 116
applied, 88, 90, 124
basic, 73, 75, 78, 79, 88, 89, 96-102, 121, 124, 136,
direction of, 15-16
Resources, 151, 154, 184
Revolution, 21, 33, 137
French, 139
Russian, 139
Rice Institute, 165
RNA, 53
Role of scientists, 68, 72, 75, 99, 168-169
Role of science, 72, 73, 74, 75, 101-102, 104, 146, 171
Role of scientists, 68, 72, 75, 99, 101-102, 141, 156, 161, 171, 174, 180-181, 182, 187-188
Role of universities, 16, 17, 87
Roman Empire, 108, 109
Royal Society (London), 103, 110
Ruhr valley, 179
Russia, 108, 142, 145, 160, 175, 176-178
Russians, 115

Saigon, 140
SALT (Strategic Arms Limitation Talks), 144
Sakharov papers, 186
Satellites, 81

Scenic Hudson Preservation case, 28
School systems, 42, 43
Science, 12, 60, 61, 62, 63, 79, 159, 161, 188
Science, 84, 104-105, 106, 107, 108, 109, 110, 111, 112, 134-136, 140, 146
applied, 156
basic, 96-102, 120, 122, 156
goals of, 72-75
role of, 72, 73, 74, 75, 101-102, 104, 146, 171
Scientific method, 98, 109, 128
Scientist, 84, 88, 95, 106, 136, 141, 143, 144, 145, 146, 156, 171, 182
attitudes of, 5, 6, 95, 104, 106, 137, 150
objectives of, 75-78
role of, 68, 72, 75, 99, 101-102, 141, 156, 161, 171, 174, 180, 181, 182, 187-188
Scientist-priest, 9, 95
Secrecy, love of, as trait of scientists, 6-7
Senate, U.S., 25
Shell Chemical Company, 62
Sierra Club, 62
Simplicity, preference for, as trait of scientists, 5-6
Social inventions, 85, 113, 123
Social responsibility, 172-173
Society for the History of Technology, 110
Solid waste, 149, 151, 154, 155
Southeast Asia, 116
South Pole, 81
South Viet Nam, 115, 116, 134
Soviet Union, 108, 142, 145, 160, 175, 176-178
Space program, 79, 89, 103, 142, 161
Sputnik, 108
SST, 29, 68, 74, 80, 184
Student role, 17-18, 77, 78, 89, 97, 108, 119
Students, 17, 18, 41, 42, 43, 44, 108, 114, 119, 120, 129, 142
Student unrest, 17, 108, 129-130, 141, 149, 157
Supersonic transport, 29, 68, 74, 80, 184
Subjective world, 104, 128
Sulfur dioxide, 148, 160, 184
Survival of man, 156, 157
Systems research approach, 37, 63, 66

Tax incentive, 179
Tax structure, 86
Tax system, 22
Technocracy, 124
Technology, 7, 10, 12, 13, 25, 27, 36, 50-51, 55, 72, 73, 74-75, 76, 84, 85, 105, 106, 113, 117, 125, 130, 140, 151, 185
Television, 41, 85, 86, 137
Test ban treaty, 86, 87, 123, 141
Texas, 149, 186
Thailand, 116
Thermal pollution, 151
Third World, 127, 136-137, 139, 140, 143, 144, 145, 162, 163, 166, 186
Tools, love of, as trait of scientists, 6
Toxins, 23, 29, 149
Transportation, 12, 23, 50, 57, 74, 80, 117, 163, 182
Trout Unlimited, 23
Truman Doctrine, 138
Two Cultures and the Scientific Revolution, The, 142

Underdeveloped countries, 136, 138, 139, 140, 142, 144, 164, 165
Unilateral disarmament, 144, 175-178
UNESCO (United Nations Educational, Scientific and Cultural Organization), 157
United Nations Conference on Trade and Development, 137
United States, 13, 21, 86, 96, 99, 127, 136, 138, 142, 144, 145, 177, 184
U.S. Geological Survey, 56
U.S. Public Health Service, 68

Universe, 92-94, 100
University, 10, 43-44, 57, 77, 87-
  90, 119, 122
basic research in, 14-15
role of, in science, 16-19, 77, 87
University Corporation for Atmospheric Research, 142
University departments, 16-18, 57-
  58, 122-123
Urban affairs and problems, 36-37,
  50, 54, 56, 66, 67, 68, 72, 74, 151,
  155, 158, 175, 183

Values, 76, 110, 156
Vatican Council, 136

Water Pollution Control Administra-
  tion, 153
*War Business: The International
  Trade in Armaments, The,* 144
Water Quality Improvement Act,
  152, 180, 183
Water pollution, 160, 179-180
Water quality standards, 151, 155
Water resources control, 22, 49
Weapons, 80, 82, 97, 115, 116, 135,
  137, 138, 139, 144, 176
Weather patterns, 23
Welfare, 38, 46-48
Whales, 65, 81
World government, 143
World survival times, 83
World War II, 99

Yale Law School, 21
Young, the, 108, 130